Energy Production and Management in the 21st Century III

WIT*PRESS*

WIT Press publishes leading books in Science and Technology.
Visit our website for the current list of titles.
www.witpress.com

WIT*eLibrary*

Home of the Transactions of the Wessex Institute.
Papers contained in this volume are archived in the WIT eLibrary in volume 222 of WIT Transactions on Ecology and the Environment (ISSN 1743-3541).
The WIT eLibrary provides the international scientific community with immediate and permanent access to individual papers presented at WIT conferences.
Visit the WIT eLibrary at www.witpress.com.

THIRD INTERNATIONAL CONFERENCE ON ENERGY PRODUCTION AND MANAGEMENT: THE QUEST FOR SUSTAINABLE ENERGY

Energy Quest 2018

Conference Chairmen

S. Syngellakis
Wessex Institute, UK

E. Magaril
Ural Federal University, Russia

International Scientific Advisory Committee

Organised by

Wessex Institute, UK

Ural Federal University, Russia

Sponsored by

International Journal of Energy Production and Management

WIT Transactions

Wessex Institute
Ashurst Lodge, Ashurst
Southampton SO40 7AA, UK

Editorial Board

Energy Production and Management in the 21st Century III

The Quest for Sustainable Energy

Editors

S. Syngellakis
Wessex Institute, UK

E. Magaril
Ural Federal University, Russia

WIT*PRESS* Southampton, Boston

Editors:

S. Syngellakis
Wessex Institute, UK

E. Magaril
Ural Federal University, Russia

Published by

WIT Press
Ashurst Lodge, Ashurst, Southampton, SO40 7AA, UK
Tel: 44 (0) 238 029 3223; Fax: 44 (0) 238 029 2853
E-Mail: witpress@witpress.com
http://www.witpress.com

For USA, Canada and Mexico

Computational Mechanics International Inc
25 Bridge Street, Billerica, MA 01821, USA
Tel: 978 667 5841; Fax: 978 667 7582
E-Mail: infousa@witpress.com
http://www.witpress.com

British Library Cataloguing-in-Publication Data

A Catalogue record for this book is available
from the British Library

ISBN: 978-1-78466-293-6
eISBN: 978-1-78466-294-3
ISSN: 1746-448X (print)
ISSN: 1743-3541 (on-line)

The texts of the papers in this volume were set individually by the authors or under their supervision. Only minor corrections to the text may have been carried out by the publisher.

Printed in Great Britain by Lightning Source, UK.

Preface

This book contains a selection of papers among those presented at the 3rd International Conference on Energy Production and Management: The Quest for Sustainable Energy, organized by the Wessex Institute of Technology, UK and the Ural Federal University, Russia.

The aim of the Conference was to discuss major challenges and complex problems in energy production and management towards economic and social development.

The rapid growth of the world population and the demand for higher living standards has led to the massive consumption of non-renewable energy sources with serious environmental consequences. While in the past many of these problems were localised, today the whole world is adversely affected, mainly due to the extensive use of fossil fuels; hence, there is an urgent need to increase the use of renewable sources of energy. At the same time, at this stage of technogenic development, it is very important to introduce innovative energy-efficient processes and technologies which enable the improvement of exploitation of traditional hydrocarbon fuels, thus increasing energy efficiency in power generation, industry and transportation and reducing environmental pollution. The solution to problems of energy sustainability requires both the consideration of national priorities and opportunities, as well as the integration of projects, research and development on the global level.

The transition from an economy based on conventional energy to one relying on renewable sources requires new scientific and technological progress related not only to energy production but also to new ways of distribution, storage and usage, aiming at mitigating the negative environmental impacts associated with conventional energy. In this context, the complexity of modern energy production and management requires a multidisciplinary approach that can take into consideration not only advances in technology but also involves the environmental, economic, social and political aspects.

The papers comprising the present volume address a wide range of issues within the research context described above. The possibility of achieving innovation through technology entrepreneurship is explored. A conceptual model is proposed for knowledge-intensive support services. The development of power industry under conditions of risk and uncertainty is studied. Reference is made to international agreements imposing targets towards achieving sustainability thus instigating national initiatives to adopt intervention strategies. The efficient use of energy is examined in the context of urban planning with particular reference to the relative energy consumption in buildings and transportation networks. A more focused investigation on energy efficiency adopts a methodology for assessing various options for public hand-drying.

Improvements in efficiency can also be achieved by adopting advanced technology. One contribution proposes a change in the operating pressure of a stabilisation process meant to ensure separation of gas and liquid products. Another proposes the use of clays for catalytic improvement of crude oils and natural gases. Organic Rankine cycle technology is proposed as a means of converting residual low grade heat into electricity. A redesigned axial compressor of a gas turbine reduces mass flow and increases specific work.

The latter part of the volume is dedicated to the generation of renewable energy. Problems interfering with the full exploitation of renewable sources at local level are discussed. Use of solar energy features in research on durable multi-layered cladding systems comprising high performance fibre-reinforced concrete, specific natural adhesives and solar cells.

Biomass is another valuable resource of renewable energy obtained from agricultural residue or waste. Related research in the volume is applied to the conversion of millet husk to liquid hydrocarbons. The last two contributions address waste-to-energy issues. Different waste flow characteristics are assessed and compared in the context of waste management in Russia. Finally, the potential risks from waste incineration are pointed out and a control methodology is proposed to avoid under-estimation of human exposure to heavy metals.

The meeting was sponsored by the WIT Transactions on Ecology and the Environment and the International Journal of Energy Production and Management. This volume is part of the WIT Transactions on Ecology and the Environment. The digital version of the papers, as well as those resulting from the previous conferences held in Yekaterinburg, Russia in 2014 and Ancona, Italy in 2016 are archived in Open Access format in the eLibrary of the Wessex Institute (https://www.witpress.com/elibrary) where they are freely available to the international community.

The editors are grateful to all authors for the quality of their contributions as well as to the members of the International Scientific Advisory Committee and other colleagues who helped to review the papers and hence ensure the quality of this volume.

The editors would like to honour the memory of Professor Carlos Brebbia by dedicating this volume to him. Professor Brebbia was the founder of the Energy Quest conference series as well as many other international meetings and research activities, co-organised by the Wessex Institute of Technology.

The Editors
The New Forest, 2018

Contents

SCALING UP THE INNOVATION PROCESS IN THE ENERGY SECTOR ON THE BASIS OF TECHNOLOGY ENTREPRENEURSHIP

LAZAR D. GITELMAN[1], MIKHAIL V. KOZHEVNIKOV[1], EVGENIY M. STARIKOV[1] & ELENA C. RADA[2]
[1]Department of Energy and Industrial Management Systems, Ural Federal University, Russia
[2]Department of Theoretical and Applied Sciences, University of Insubria, Italy

ABSTRACT

The liberalization of the energy sector, the spreading adoption of digital information and telecommunication technologies, decentralization, and expansion of RES-based distributed generation provide broader development opportunities for technology entrepreneurship in the industry. Nevertheless, the energy sector remains one of the "least entrepreneurial" industries of the global economy – both by number of corporate innovations and by number of independent startups. Having summarized a corpus of theoretical research and the results of a survey of energy experts, the authors were able to describe fields of applications and key barriers to the development of technology entrepreneurship in the industry and at a corporate level. On the basis of an analysis of the experience of US and British energy companies, the authors propose typical models of technology entrepreneurship and schemes of resource interaction among energy companies, consumers of new products and services, universities, developers and investors in the course of implementation of breakthrough projects. The results of the study and the authors' recommendations can be used by energy companies creating corporate mechanisms of technology entrepreneurship, as well as by government agencies and universities for the development of the innovation infrastructure of the energy sector.

Keywords: technology entrepreneurship, startup, smart energy, digitalization, technology modernization, innovation infrastructure.

1 INTRODUCTION

The authors define technology entrepreneurship as a high-risk activity of implementing projects that are aimed at creating new technologies on the basis of systems of continuous search for emerging opportunities, transformation of the newest scientific and technical developments into original business ideas and accumulation of dispersed resources that are required for bringing innovative products and services to the market. Created technologies may be represented as supporting innovations in various fields (management, energy production, business models), and disruptive, fundamentally changing the entire market landscape [1].

Interest in this phenomenon peaks at the moments of radical technological shifts. As shown in [2], [3], technology entrepreneurship plays a support role at the growth stages of an innovation's life cycle that is represented by Gartner's hype cycle.

In the Technology Trigger phase, the task of technology entrepreneurship is to boost research and the launch of startups, to forge partnerships and strategic alliances. In the phases of secondary growth and robust productivity (Slope of Enlightenment and Plateau of Productivity), technology entrepreneurship has an adjustment function of helping organizations to chisel technological solutions and complete the market adaptation to new products.

Apart from that, big companies and universities, who are the main "consumers" of technology entrepreneurship, view it as a source of organizational flexibility and a fuel for corporate innovation spirit as technology entrepreneurs often operate as creative local teams that focus on various engineering and marketing aspects of a technological solution [4], [5].

WIT Transactions on Ecology and the Environment, Vol 222, © 2019 WIT Press
www.witpress.com, ISSN 1743-3541 (on-line)
doi:10.2495/EQ180011

In the energy sector that is currently going through radical transformations resulting from broad smartization, energy market reconfiguration, the expansion of RES-based distributed generation and smart grids, electric transport and energy storage and accumulation solutions, technology entrepreneurship has been developing at a slower pace as compared to other industries. It is estimated that expenditures for R-and-D in the energy sector have been stagnating since 2012. In 2015, investment in technological innovations globally amounted to 67bn dollars, which is only 4% of the global energy investment [6]. Moreover, in 2015–2016, investment to modernize the grid and generating facilities showed a slight downward trend – for the first time in 15 years. By contrast, in the IT industry, nearly 40% of total investment was directed towards the development of innovative solutions [7].

Nevertheless, a number of successful examples show that technology entrepreneurship could be an effective and *indispensable* mechanism for boosting innovation – both at the level of individual energy companies and at an industry level. In this study, we will identify typical barriers to technology entrepreneurship development and formulate models and schemes of resource interaction in the process of implementing breakthrough projects among energy companies, consumers of new products and service, universities, developers, and investors. In conclusion, we will give recommendations for heads of energy companies and government bodies as regards the creation of effective infrastructure that supports the development of this area.

2 CHARACTERISTICS OF TECHNOLOGY ENTREPRENEURSHIP

Technology entrepreneurship has been on scholars' radar since the 1980s. The scientific community was, however, more interested in the economic rather than process-related aspects: for example, the issue of value migration, the ability of technology to be sold as a standalone product, the structure of the market for technology [8]–[11]. In the early 2000s, Shane and Venkataraman attempted to provide a more universal understanding of technology entrepreneurship by defining it as an activity aimed at "assembling organizational resources and technical systems, and the strategies used by entrepreneurial firms to pursue opportunities" [12]. Later, Bailetti summarized the works of many of his earlier counterparts and defined technology entrepreneurship as an investment in a project that assembles and deploys specialized individuals and heterogeneous assets advancing scientific and technological knowledge to create and capture value for the firm [13].

At the forefront of technology entrepreneurship there is a search for new, breakthrough and research-intensive technologies (a device, a computer program, or a unique service) that guarantee a firm or a business eco-system a long-lasting competitive edge and that have a value to the scientific community thanks to their substantial hidden potential as a research subject.

As the end product requires a large volume of new scientific knowledge that can hardly be generated by a single structure, be it a high-tech company, a startup or a university, those involved in the process of technology entrepreneurship have to engage in close cooperation.

Such cooperative groups might include:

- scientists/ research teams/universities;
- independent entrepreneurs/ business-structures/investors;
- enthusiasts/ pro-active employees.

The makeup of participants, the focus of research and development projects, the degree of novelty and continuity in technologies being created serve as the classification characteristics for technology entrepreneurship. In our view, the characteristics could be used for identifying the typical forms of technology entrepreneurship that differ in terms of stakeholders involved, nature of their interaction and market goals (Table 1).

Various forms of technology entrepreneurship could appear in various business models. Gabriel and Kirkwood [14] used a survey of over 100 respondents in 28 countries to derive four business models in technology entrepreneurship that are typical of the energy sector and complex industries. The models are termed "Consultants", "Distributors", "Integrators", and "Technology Owners" (Fig. 1).

The main differences between the above business models lie in the comprehensiveness of the product offering and their focus on one or several stages of the lifecycle of the innovative process. "Consultants" as the most basic type of technology business tend to provide service rather than manufacture tangible products. They offer the most versatile range of auxiliary services within the lifecycle of an innovation: market analysis and forecasting, product and systems design, expert analysis of research-intensive solutions, consulting. "Distributors" act as technology intermediaries between the developers and buyers of new systems.

Table 1: Forms of technology entrepreneurship.

Characteristics	Short description	Details
Participant makeup	Firm	The firm only uses its own resources to implement technology entrepreneurship projects
	Firm and university (also individual scientists or research teams)	The firm cooperates with the university (as a whole or individual scientists/ research teams) and participates in experience exchange events
	Startup (in cooperation with investors)	A group of enthusiasts sets up a small technology firm that uses investors' funds for growth and product development
	University (in cooperation with consultants and investors)	The university's intellectual property is commercialized with the assistance of investors and business consultants
Focus of research	Corporate research undertaken in-house	Innovations are developed and utilized within the company
	Externally commissioned research	The company makes an R&D request that is delivered by market participants
	University spinoff	The university conducts a series of studies the results of which are patented or form a foundation for a small innovative business in which the university is the principal shareholder
	"Garage" R&D	Individual enthusiasts or groups of enthusiasts who work on technologies of interest (e.g. first PCs assembled in California)
Relation to existing technology	New technologies	Technological solutions that are completely new to the market (breakthrough innovations)
	Existing technology modernized	New technology functionality discovered, or technology properties transformed to improve work of systems
	Mix of existing and new technologies	When used together, they acquire new properties and become a new product (e.g. pilotless automobiles)

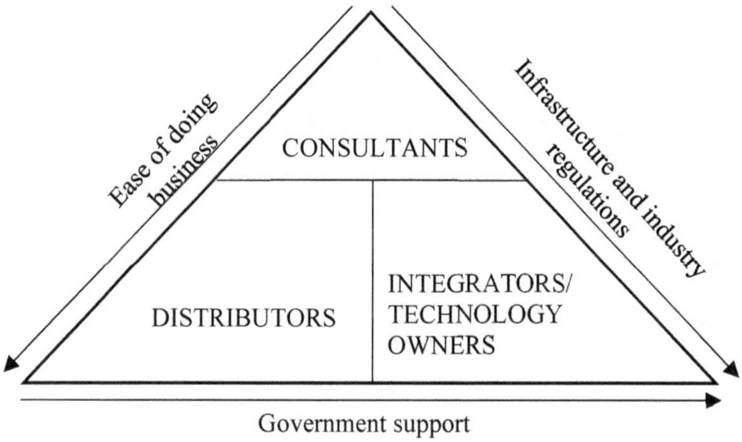

Figure 1: Triangular matrix of business models for technology entrepreneurship [14].

They specialize in the assembly of a turn-key system from different components to meet the customer's requirements. "Integrators" independently design, construct and produce complex systems while providing ancillary logistics, information and technical services. Finally, "technology owners" focus on inventing fundamentally new technological solutions and principles that can subsequently serve as the foundation for a unique production system.

The presented forms of technology entrepreneurship exist in the energy sector. There are, however, specific barriers and corporate inertia of energy company that could hamper the process of innovation. We shall take a closer look at these aspects further in the text.

3 ANALYSIS OF GLOBAL EXPERIENCE OF TECHNOLOGY ENTREPRENEURSHIP IN THE ENERGY SECTOR

Globally, technology entrepreneurship in the energy sector has been predominantly developing in two directions: adaptation of renewable energy sources to satisfy the requirements of mass consumption (e.g. creation of various services that enable consumers to integrate into the energy market) and creation of energy efficient and environmentally friendly technological solutions for energy systems. Technology entrepreneurship could, however, find a broader application. The authors have studied over 30 cases of technology entrepreneurship of various scales and composition (Table 2) that demonstrate the wide applicability of this tool.

Table 2 presents some of the most interesting cases of technology entrepreneurship that indicate the diversity of participants, their interdependence and the synergistic effect that arises from their cooperation and resource exchange. A separate category of cases is made up of companies that not only embrace innovative technologies, but also adopt new business models.

For example, in 2016 Siemens and LO3 Energy teamed up to build a local smart grid in Brooklyn. LO3's system is intended to let "prosumers" buy and sell energy generated from rooftop solar panels to their neighbors. LO3 users install a high-resolution meter, which tracks energy usage at specific times of day, or an app that lets them set buy and sell requests for specific kinds of electricity, such as solar or wind. LO3 gets revenue from the resulting transactions, and also wants to penetrate the bigger utility market, potentially by connecting microgrids to major utility operations. LO3 also considers using blockchain and the smart

Table 2: Technology entrepreneurship cases in energy sector [15]–[25].

Name of energy company/ organization/project	**Technology being introduced**	**Stakeholders**
Hangzhou Zhongheng Electric (China)	Sensors concealed within workers' uniforms can analyze their emotions and help increase their productivity through timely job rotations	Energy company
Smart Grid Project in Belgorod (Russia)	"Smart" energy consumption measurement based on automated system of commercial electricity measurement using Neuron smart meters	Energy company and state
Machine learning for business process management at General Electric (USA)	Corporate university that specializes in training the company staff in machine learning technology and building learning digital twins that help predict equipment failures	Energy company and university
Carnegie Mellon University (USA)	Energy consumption management system for companies that incorporates a smart control board for monitoring energy costs and reducing them at peak loads	University
Aquion Energy (USA)	Aqueous Hybrid Ion Battery that uses a water-based sodium sulfate electrolyte for energy storage	University and startup
Shell (USA)	The oil company has invested in a startup that sells solar energy for a flat monthly rate	Energy company and startup
Moya Power (UK)	Big plastic sheets are installed along railways and in tunnels. The sheets generate electricity by capturing air streams.	Startup
Construction of smart grids in Australia	A project to build distributed energy generation installations that are connected to a smart grid for energy redistribution. Solar panels and batteries rolled out to 50,000 homes will be pulled together into a single virtual power plant. Project jointly implemented by an investment fund, the government and Tesla.	Energy company (Tesla) and investment fund and government

contracts at the heart of second-generation blockchains like Ethereum to automate peer-to-peer energy transactions.

Amid falling oil prices, Royal Dutch Shell started to invest in alternative energy companies. Its board of directors made a strategic decision to invest 1 billion dollars per year in renewables, and technologies of smart construction and energy management. One of its pilot projects is Inspire Energy Holdings that offers wind and solar energy at a flat subscription price. Inspire also provides smart home and energy management services via a mobile app.

Finally, there is Nest, a startup backed by telecommunications giant Google. Nest positions itself as a provider of smart services for households – from thermostats and smart energy meters to home security systems to security cameras and fire alarm systems. What makes Nest's business model peculiar is that all the devices it has ever sold are bundled into a single network using machine learning. At the moment, Nest is a service aggregator that manages over 50,000 thermostats in California and builds its economy on the network effect: the more households are connected to the platform, the more lucrative the company is.

It is mainly developed countries that can boast of successes in technology entrepreneurship in the energy sector. By contrast, developing countries often have serious issues with institutionalizing industry-specific innovations because the structure of the energy market, its flexibility and non-discriminatory access to the market infrastructure constitute the key factors that shape a comfortable environment for innovation.

Having analyzed the peculiarities of energy markets in developed and developing countries, the authors have identified the main barriers that act as constraints technology entrepreneurship development.

A rigid legal framework that leaves companies limited or no room for independent decision making. All too often companies even have to coordinate their investment policy with the authorities, which minimizes their capacity for responding promptly to changing market requirements and for developing new technologies.

Market inertia and weak competition. Any energy market is an artificial construct that consists of naturally monopolistic companies (electric power transmission, operations control) and competitive ones (generation, sales, repair and maintenance, engineering, none-core activities) as well as open wholesale and retail markets for electric power. The paradox is that market participants that directly interact with consumers (for example, power supply companies) find themselves at the very end of the value chain and de facto have no freedom of strategic maneuver.

Closing the gap between the current expectations of consumers and what energy companies have to offer would only be possible if the vertical hierarchy became weaker and service "in-fill layers" between centralized and distributed generation, the distribution and sales sectors. This would have a possible effect on the diffusion of innovations and boost competition in the market [26], [27].

Conservatism of market participants. The corporate culture of many energy companies often features a specific mode of behavior that does not welcome fast change. As a result, energy companies passively react to changes in the external environment and managers tend to make decisions reactively. New technologies that are perceived by the market as breakthrough ones are introduced slowly and sporadically. The search for innovations is not systemic and is not part of the company's strategy.

It is appropriate to quote energy systems expert Lee Bryant as saying in his blog that "the irony of companies that talk about smart grids, distributed networks, resilient supply and smart IoT devices, is that they operate internally using email, PowerPoint and endless meetings and conference calls. To succeed in connected markets with connected products, you need to be a connected company" [28]. The expert thus stresses that the success of energy companies depends on their readiness to employ technology and, consequently, on the availability of the organizational paradigm, flexible organizational systems and culture that encourage the process of innovation, rather than on their awareness of the technological context.

The above arguments are additionally backed by the findings of a survey that the authors conducted to ask experts at regional energy companies about their attitude to technology entrepreneurship. According to the survey, 56% of the interviewees do not implement technology entrepreneurship project, but admit it is today's big trend. An overwhelming majority of the respondents said that in most cases the only way to encourage the process of innovation is by declaring support for ideas and concepts proposed by employees rather than through systemic search for innovations (Fig. 2,3). Among the most promising trends in technology innovation, staff members of energy companies name advances in core production technologies, information systems, new services for consumers, engineering and repair services (Fig. 4).

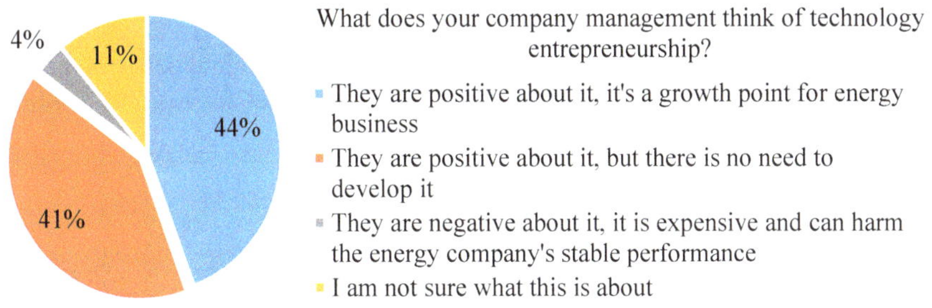

Figure 2: Attitude of leaders of energy companies to technology entrepreneurship.

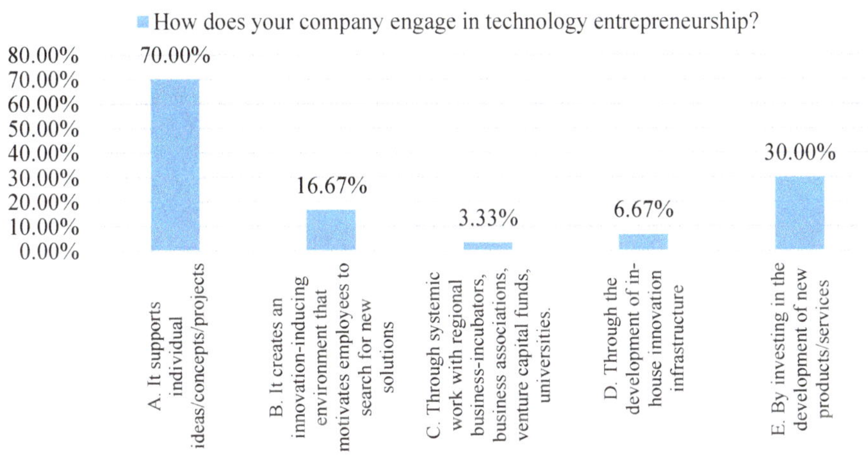

Figure 3: Ways of implementing technology entrepreneurship projects.

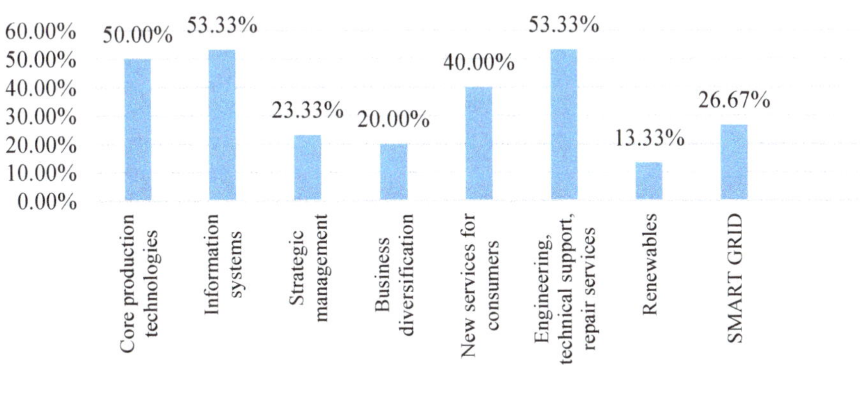

Figure 4: Promising trends in technology entrepreneurship.

4 MODELS OF TECHNOLOGY ENTREPRENEURSHIP IN THE ENERGY SECTOR

Having conducted an analysis of the meaningful traits of technology entrepreneurship, actual cases and expert opinions, the authors were able to identify typical models that demonstrate mechanisms of introducing breakthrough innovative technologies, their application areas and capabilities. Table 3 is a generalized summary of the models.

Table 3: Models of technology entrepreneurship in the energy sector.

Model	Description	Pros	Cons
Government subsidies for projects in major energy companies	Major infrastructure projects are implemented with support from government	– Major projects can be implemented that the company would not be able to undertake on its own – Some risks are covered; the government acts as a "lifesaver" – Access to all available infrastructure that makes it easier to implement the project	– Red tape in decision making affects project deadlines – Increased control on the part of government
Energy company's investment in in-house R&D	Intellectual and financial resources are mobilized within the company and used as the basis for creation of innovations	– Technologies can be tailored to specific tasks of the energy company – Absence of intermediaries or other parties that have an influence on the process and demand performance results – Proprietary information can be kept secret – Development of corporate R&D and positive changes in the organizational culture of the business	– Higher risks as the company fully funds projects – As a result, big capital-intensive projects become unfeasible – A project is at risk of not getting a comprehensive implementation because the company may not have the necessary expertise of resources
Commercialization of university R&D by energy companies	University R&D products are transformed into marketable products	– An opportunity to convert technologies that business may not have considered profitable or promising into innovations – University has an abundance of intellectual resources that could be utilized in current and subsequent projects – An opportunity to perfect mechanisms of knowledge and technology transfer, cooperation and networking – Access to new staff and competencies	– Lack of managerial experience requires the presence of consultants on the team – Few companies have the potential to be commercially successful
Launch of startups as a standalone or daughter business	Investment is attracted for startup launch	– Absence of red tape and complete freedom of creation – Broad opportunities for testing technologies/products/services being created	– Very high risk of project failure or the failure to find a working business model – Involvement of professional consultants is required to help the startup survive through growth stages

Government subsidies for projects in major energy companies. The majority of projects in the energy sector require huge investment. In most cases, it is the government that acts as the guarantor as it is ready to invest overhaul projects in critical infrastructure by taking on the biggest share of obligations to absorb financial and organizational risks.

Such a model of implementing national and regional energy projects is typical of strategic initiatives and can be used for smart city development projects and smart energy systems, distributed generation clusters, pilot generation installations employing emerging technologies. From the point of view of technology entrepreneurship, energy companies can be interested in the model because it opens up opportunities for the development of innovative industrial infrastructure but can hardly help it build its own entrepreneurial competences.

An energy company's investment in in-house R&D. Mobilizing intellectual and financial resources within the energy company and creating innovations on the basis of emerging growth points is another popular model. The majority of big corporations that hold stable positions in national and global markets opt for this model of entrepreneurial behavior. To execute the model, a standalone corporate unit can be set up that takes responsibility for the process of innovation and performs context analysis, conducts R&D activities and product testing. If the tests prove successful, the energy company may turn this activity into a separate business. The model is typical of projects that are aimed at business process optimization or the improvement of technologies in use.

Commercialization of university R&D with the help of external financing. The model is essentially about building close cooperation with universities for conducting research, acquiring new knowledge and unique R&D achievements. The model could be utilized by energy companies in local projects of technology modernization but are most frequently employed when creating breakthrough and fundamentally new technologies. One vivid example is numerous partnerships of leading players in the nuclear market to design Gen IV nuclear reactors.

Launch of startups. The model is picked by both enthusiasts who are prepared to seek financing for their ideas, and by companies whose core business lies outside the energy sector, but who want to diversity into the energy market. The range of products that can be developed within the framework of the model is immensely broad and can include information applications for consumers, new equipment, systems of energy process monitoring, components for smart energy systems. The startup model is a service one from the energy market perspective and has a huge potential for creating specialized platforms of technology entrepreneurship.

Relations between the models are depicted in Fig. 5. This conceptual scheme integrates the main connections and functions of the process participants, directions of their resource and information interactions. The scheme could be used for the development of technology entrepreneurship infrastructure.

5 CONCLUSION

The presented models can be employed by heads of energy companies as an effective tool for ramping up the process innovation, creating long-lasting competitive advantages, building customer relationships, and supporting complex projects of technology modernization. Given the complex attitudes to technology entrepreneurship in the energy sector, one needs to define baseline conditions the absence of which makes it inexpedient to embrace technology entrepreneurship.

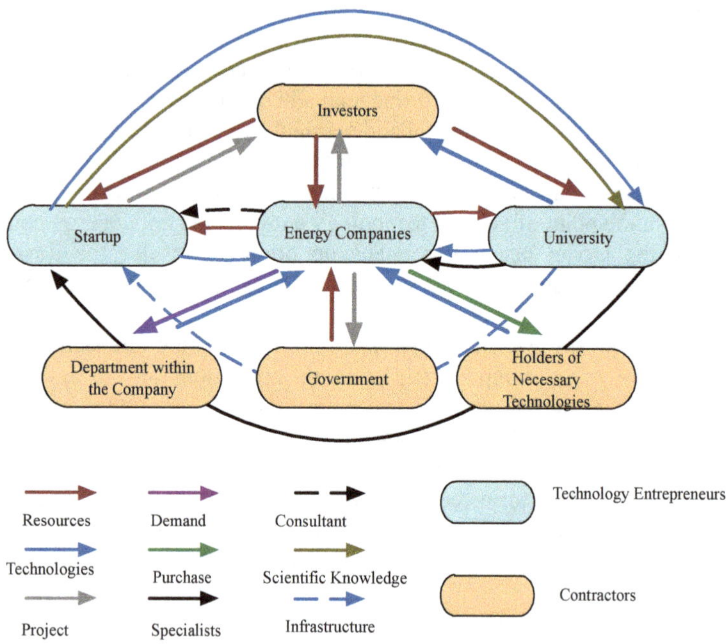

Figure 5: Integrated scheme of technology entrepreneurship models in energy sector.

Technology entrepreneurship is good for implementing *breakthrough ideas* whether it comes to technology, products, services or management systems. Otherwise one may not be able to recoup the costs and the risks will make not worth the effort. The energy company must have a clear idea of its position and resources in the chain of breakthrough innovations that correspond to trends in the regional, national, or global energy policies – depending on the size of the energy business.

Technology entrepreneurship is difficult to do *without R&D and close cooperation with science*. That's why universities and energy companies that focus on innovation establish special closely interlocking structures for knowledge transfer that are designed to monitor the latest achievements in science and technology and exchange best practices.

Technology entrepreneurship cannot be done without building teams that continuously search for new ideas, methods of their implementation and create a flow of innovation – a kind of bloodstream in the energy company. It becomes vital to create *appropriate organizational culture* without which not a single stage of technology entrepreneurship can be executed.

These ideas are the subject of further research and require additional elaboration. The authors' own experience [29], [30] of forming breakthrough teams in energy companies and developing smart partnerships of energy business, science and education has been highly successful and provides a convincing proof of the practical relevance of technology entrepreneurship as an instrument of responding to industry challenges.

ACKNOWLEDGEMENT

The work was supported by Act 211 of the Government of the Russian Federation, contract № 02.A03.21.0006.

REFERENCES

[1] Christensen, C.M., *The Innovator's Dilemma: When New Technologies Cause Great Firms to Fail*, Harvard Business School Press: Boston, 1997.

[2] Linden, A. & Fenn, J., Understanding Gartner's Hype Cycles, Strategic Analysis Report. www.bus.umich.edu/KresgePublic/Journals/Gartner/research/115200/115274/115274.pdf. Accessed on: 30 Jul. 2018.

[3] Dedehayir, O. & Steinert, M., The hype cycle model: A review and future directions. *Technological Forecasting and Social Change*, **108**(C), pp. 28–41, 2018. DOI: 10.1016/j.techfore.2016.04.005.

[4] Abdullah, S., Musa, C.I. & Azis, M., The effect of organizational culture on entrepreneurship characteristics and competitive advantage of small and medium catering enterprises in Makassar. *International Review of Management and Marketing*, **7**(2), pp. 409–414, 2017.

[5] Muñoz, R.M., de Pablo, J.D.S., Peña, I. & Salinero, Y., The effects of technology entrepreneurship on customers and society: A case study of a Spanish pharmaceutical distribution company. *Frontiers in Psychology*, **7**, pp. 1–13, 2016. DOI: 10.3389/fpsyg.2016.00978.

[6] International Energy Agency. World Energy Investment 2017. www.iea.org/publications/wei2017/. Accessed on: 30 Jul. 2018.

[7] Mas, M., Fernández de Guevara, J., Robledo, J.C. & López-Cobo, M., The 2017 PREDICT Key Facts Report. An Analysis of ICT R&D in the EU and Beyond, 2017. http://publications.jrc.ec.europa.eu/repository/bitstream/JRC106658/kjna28594enn.pdf. Accessed on: 30 Jul. 2018.

[8] Cooper, Arnold C. & Komives, John L. (eds), Technical Entrepreneurship: A Symposium. Center for Venture Management, Milwaukee, WI, 1972.

[9] Florida, R. & Kenney, M., Venture capital and high technology entrepreneurship. *Journal of Business Venturing*, **3**(4), pp. 301–319, 1988. DOI: 10.1016/0883-9026(88)90011-0.

[10] Bahrami, H. & Evans, S., Flexible re-cycling and high-technology entrepreneurship. *California Management Review*, **37**(3), pp. 62–89, 1995. DOI: 10.2307/41165799.

[11] Kenney, M. & von Burg, U., Technology, entrepreneurship and path dependence: Industrial clustering in Silicon Valley and Route 128. *Industrial and Corporate Change*, **8**(1), pp. 67–103, 1999. DOI: 10.1093/icc/8.1.67.

[12] Shane, S. & Venkataraman, S., Guest editor's introduction to the special issue on technology entrepreneurship. *Research Policy*, **32**(2), pp. 181–184, 2004. DOI: 10.1016/s0048-7333(02)00104-x.

[13] Bailetti, T., Technology Entrepreneurship: Overview, Definition, and Distinctive Aspects, Technology Innovation Management Review. https://timreview.ca/article/520. Accessed on: 30 Jul. 2018.

[14] Gabriel, C.-A. & Kirkwood, J., Business models for model businesses: Lessons from renewable energy entrepreneurs in developing countries. *Energy Policy*, **95**, pp. 336–346, 2016. DOI: 10.1016/j.enpol.2016.05.006.

[15] Morris, D., Siemens to Invest in Blockchain-Based Smart Grid Builder LO3, Fortune. http://fortune.com/2017/12/24/siemens-blockchain-lo3. Accessed on: 30 Jul. 2018.

[16] South China Morning Post, "Forget the Facebook leak": China is mining data directly from workers' brains on an industrial scale. www.scmp.com/news/china/society/article/2143899/forget-facebook-leak-china-mining-data-directly-workers-brains. Accessed on: 30 Jul. 2018.

[17] Nikolaev, M.V., Ovsyannikov, A.Yu. & Marchinsky, I.Yu., Experience of using smart networks in Russia. Technology Smartgrid. Young Scientist, 28(2), pp. 63–68, 2016, https://moluch.ru/archive/132/37010. Accessed on: 30 Jul. 2018.

[18] Woyke, E., General Electric Builds an AI Workforce, MIT Technology Review, www.technologyreview.com/s/607962/general-electric-builds-an-ai-workforce. Accessed on: 30 Jul. 2018.

[19] Carnegie Melon University, Press Release: Carnegie Mellon Researchers Develop Dashboard to Help Reduce Energy Use in Office Buildings. www.cmu.edu/news/stories/archives/2014/april/april7_workplacedashboard.html. Accessed on: 30 Jul. 2018.

[20] Allan, R., Researchers look to materials to Improve Micro-Grid Renewable Energy Storage. www.electronicdesign.com/power/researchers-look-materials-improve-micro-grid-renewable-energy-storage. Accessed on: 30 Jul. 2018.

[21] Eckhouse, B. & Gilblom, K., Shell's Pivot to Renewables Sharpens with California Deal, Bloomberg. www.bloomberg.com/news/articles/2018-02-14/shell-makes-another-bet-on-renewable-energy-with-inspire-deal. Accessed on: 30 Jul. 2018.

[22] Peteranderl, S., Crossrail will generate electricity using the wind created by trains, Wired. www.wired.co.uk/article/city-wind-farms-energy-crossrail-london-renewable-energy. Accessed on: 30 Jul. 2018.

[23] Lambert, F., Tesla is installing Powerwalls and solar power on 50,000 homes to create biggest virtual power plant in the world, Electrek. https://electrek.co/2018/02/04/tesla-powerwall-solar-virtual-power-plant. Accessed on: 30 Jul. 2018.

[24] Peters, A., These Oil Industry Workers Now Teach Their Colleagues How To Install Solar, Fast Company. www.fastcompany.com/40506161/these-former-oil-industry-workers-now-teach-their-old-colleagues-how-to-install-solar?utm_source=feedly& utm_medium=webfeeds. Accessed on: 30 Jul. 2018.

[25] Buchmann, M., Digital Platforms in the Energy Sector: Definition and First Applications www.theenergycollective.com/enerquire/2410260/digital-platforms-energy-sector-definition-first-applications. Accessed on: 30 Jul. 2018.

[26] Pelzer, A., Lombardi, P., Arendarski, B. & Komarnicki, P., An innovative energy management system for the integration of volatile energy into industrial processes. *International Journal of Energy Production and Management*, **1**(4), pp. 339–348, 2016. DOI: 10.2495/eq-v1-n4-339-348.

[27] Weiller, C.M. & Pollitt, M.G., Platform markets and energy services. Working Paper, 2013. www.eprg.group.cam.ac.uk/wp-content/uploads/2013/12/1334-PDF.pdf. Accessed on: 30 Jul. 2018.

[28] Bryant, L., Can utility companies make the transition to energy platforms? https://postshift.com/can-utility-companies-make-the-transition-to-energy-platforms/. Accessed on: 30 Jul. 2018.

[29] Gitelman, L.D. & Kozhevnikov, M.V., A paradigm of managerial education for a technological breakthrough in the economy [in Russian]. *Economy of Region*, **14**(2), pp. 433–449, 2018. DOI: 10.17059/2018-2-8.

[30] Gitelman, L.D., Gavrilova, T.B., Kozhevnikov, M.V. & Starikov, E.M., Ensuring resilience and agility of complex organizational-technical systems. *International Journal of Design & Nature and Ecodynamics*, **13**(2), pp. 208–220, 2018. DOI: 10.2495/DNE-V13-N2-208-220.

A TRANSITION TO KNOWLEDGE-INTENSIVE SERVICE ACTIVITIES IN POWER INDUSTRY: A THEORETICAL FRAMEWORK

MIKHAIL V. KOZHEVNIKOV
Department of Energy and Industrial Management Systems, Ural Federal University, Russia

ABSTRACT
The current smart energy transition is accompanied by major transformations in the aspects of technology, market, and organization. The intensive adoption of digital technology in energy production, the rapid spread of distributed generation, micro-grids and energy storage and accumulation solutions, and an ever-increasing integration between the energy sector and other critical infrastructure sectors have brought about fundament changes in the relationships between energy companies and the market. This creates a strong demand for versatile support services for these processes. The article makes an attempt to construct a conceptual framework for and to outline the tasks and goals of knowledge-intensive services in the new energy industry. The author provides justification for the key properties of knowledge-intensive services that include adaptability, the capability of self-tuning, a wide use of information and telecommunication technologies, a tendency to use a pro-active approach to systems and process management, and the availability of staff with outstanding qualifications. The author has developed a conceptual model of knowledge-intensive services in the energy sector that makes it possible to group the service segments by the dimensions of "energy market optimization and customer relations", "asset lifecycle management", "organizational flexibility of energy market agents". Recommendations are given as to how to develop knowledge-intensive service activities in energy companies, including through the creation of corporate platforms and smart partnerships with universities and science.
Keywords: knowledge-intensive services, information and telecommunication technology, smart energy, technology modernization, critical infrastructure, platform, energy company, conceptual model.

1 INTRODUCTION

Over the past few decades, pivotal structural shifts have occurred in the global economic landscape. For a long time, the production of tangible goods was given priority in the economic reproduction of the country, while services were considered to play a secondary role. In other words, it is the commodity composition of industrial output that ensures a stable position of the country in the system of the international division of labor and its economic security, i.e., its resilience during geopolitical crises, while the service sector follows the development of industry and supports the efficiency of industrial production in the broad sense of the word.

Nevertheless, the transition from an industrial economy to a knowledge economy and on to a new digital economy is accompanied with paradigm shifts in corporate business models – from the mass production of standardized tangible products to a diversified range of services that give an additional value to the product offer. As a result, the tertiary sector has started to have a defining impact on the economic sustainability of countries and regions. The impact is determined by globalization, a high level of entrepreneurial activity in service markets [1], [2], a stable growth in the consumption of services as a distinct product [3], [4], unprecedented employment rates in service businesses [4]–[6], and a higher degree of innovation as compared to traditional industries [7]–[10]. Annual growth rates for value added in services exceed those in manufacturing in the overwhelming majority of countries

www.witpress.com, ISSN 1743-3541 (on-line)
doi:10.2495/EQ180021

(Fig. 1). In some countries (Great Britain, France, US), the share of the tertiary sector in GDP is nearly 80%.

However, the boundaries of the definition of "tertiary sector" that was introduced by Fisher [11] and later refined by Clark [12] are very conditional. International industrial classifications such as NACE, ISIC or NAICS categorize a wide range of various activities as the tertiary sector. This does not fully reflect the profound transformation of the service industry. A close integration of services and manufacturing, rapid technological modernization of economies, sweeping digitization and personalization, the processes of service externalization result in the sprouting of new service segments, with so-called "knowledge-intensive service" that serves as a "facilitator of innovations" being the most interesting one.

Research in the area began to emerge in the 1990s and has largely been of general nature. A few cases were dedicated to certain aspects of knowledge-intensive services in the IT industry, the oil and gas sector, health care and pharmaceuticals, and agriculture [13]–[17]. Meanwhile, it is the sectoral affiliation that fills knowledge-intensive service with specific and meaningful tasks and determines its organizational features. We shall consider these features using the case of the energy sector as one of the most complex industries that is going through dramatic transformations affecting both production and technology and relations between energy companies and consumers against the backdrop of a new industrial revolution.

2 THEORETICAL BACKGROUND FOR STUDIES OF KNOWLEDGE-INTENSIVE SERVICE

Knowledge-intensive services have been widely adopted in hi-tech industries such as pharmaceutical and medical devices manufacturing, electric power generation and the oil and gas industry, machinery manufacturing, computer and peripheral equipment manufacturing, wired and wireless telecommunications; semi-conductor and other electronic component manufacturing; the aerospace industry; geolocation services; engineering and consulting; electric equipment manufacturing; chemical manufacturing; industrial design [14]. The most authoritative researchers of knowledge-intensive service [18]–[21] define it as an independent segment of specialized business services that have a high degree of the information and intellectual elements in their value. Knowledge-intensive services can, for example, cover software development, management consulting, scientific and technical

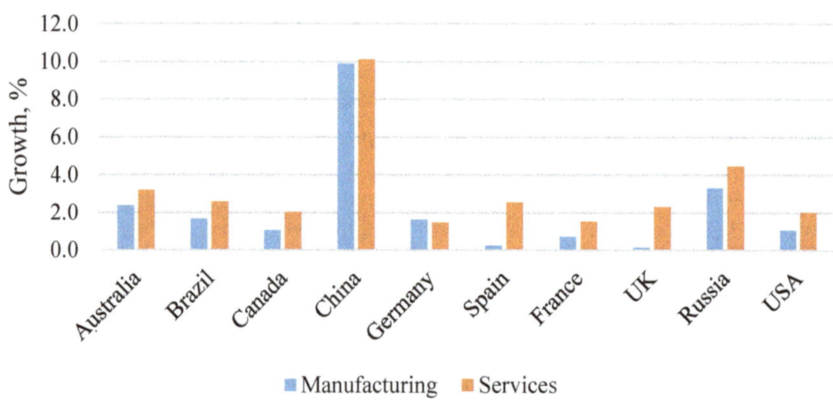

Figure 1: Growth rates for added value in manufacturing and service in 2000–2017.

evaluation; business audit; marketing analysis and foresight; complex machinery repair; engineering, data storage and analysis, telecommunication services; personnel training. Kox and Rubalcaba [19] and Wong and He [22] note that knowledge-intensive services are mainly traded in business-to-business transactions that involve two types of parties:

- developers of technological know-how, scientific studies, intellectual property that can be converted into various marketable forms, and
- business consumers of "advanced knowledge" who seek to become leaders of innovation in the market by performing continuous content analysis and attracting experts with new competencies.

The process of producing knowledge-intensive services depends on: 1) the entrepreneurial activity of the service company; 2) its institutional status; 3) technology cooperation with knowledge transfer centers; 4) the degree of industrial/corporate dispersion. Unlike traditional services that can be delivered in a standardized way in standalone business units or branches of the company, knowledge-intensive services need to be organizationally distributed.

The interdisciplinary nature of the process of producing knowledge-intensive services makes it impossible to keep the service function within a specific organizational cell because intangible and tacit knowledge needs to be exchanged constantly, both internally and externally.

This is what makes knowledge-intensive services a kind of interface between providers and clients, a zone when advanced knowledge is concentrated and solutions to the most complex and extremely specific problems are found in the most efficient, timely and comfortable way. The research element becomes an essential part of knowledge-intensive services [23].

The process of knowledge-intensive service consumption is affected by the so-called "socio-technological paradox": customers simultaneously have negative and positive expectations in relations to one and the same effect (labor productivity, market expansion, business process optimization, innovative development). At the same time, the customer usually wants to be present at the delivery of the service (to monitor it, get relevant information and even provide assistance to the contractor). There occurs the co-opetition effect – a combination of cooperation and competition between the parties involved in service processes for the right to select the best way of service delivery, which motivates contractors to constantly work on the quality of service processes [24]–[26]. At the same time, customers' engagement in service delivery gives businesses an opportunity to create platforms that scan the market for new preferences, rapidly design new service, rapidly tests them and brings to the market and promptly collects feedback and adds improvements to the ultimate service [27].

The rise of knowledge-intensive services has been particularly strong in the areas where there is an abundance of emerging technologies (Table 1) and related digital eco-systems. Such segments set forth new requirements to services as a support function for production and business processes. A digital eco-system alone does not turn out a product; it is created as a pre-emptive response to disruptive events in order to minimize the adverse effect of market, social or technological risks [28]. For new production technologies to function effectively, they need to be surrounded by a network of adaptive self-tuning services that make it possible to monitor the parameters of the system on a real time basis as well as to predict its behavior in the future.

The fundamental requirement that the provider of a knowledge-intensive service needs to meet is that the competencies of the service personnel must be more diversified than in traditional manufacturing firms. Gölpek [7] observes that service personnel must have a

Table 1: New manufacturing technologies and ancillary knowledge-intensive services.

Technology	Traditional technology and equipment	New technology and equipment	Types of knowledge-intensive services
Technology for organization and management	Systems engineering approaches, LEAN	Smart or cloud manufacturing, digital firms	Data storage, cyber security, digital consulting, knowledge management
IT-systems supporting product lifecycle	CAD, CAE, CAM / PDM	CAx for additive technologies, M2M, cloud technologies	Spatial modeling of complex objects; maintenance services for smart industrial management systems
Fabrication technology	Machinery manufacturing	Additive manufacturing	3D prototyping, stereolithography, selective laser melting
Automation equipment and technology	Relays, switches, sensors, power electronics	Industrial robots	Computer simulator, service robots, CNC machining
Advanced materials	Metals, plastics	Composite materials with programmable properties	Materials testing
Logistics	Push-and-Pull Scheduling	Multi-agent logistic systems	Virtual planning
Transport	Mass production of transportation vehicles power by alternative fuels	Smart electric transport systems	Adaptive traffic control, smart parking; services supporting the trading of energy between electric transport and energy market

thorough understanding of the functional peculiarities of new manufacturing technologies through all stages of their lifespan. Such technologies usually require constant human-machine interaction by means of special service interfaces, which means round-the-clock supervision of the operation of complex manufacturing systems and customers who use new manufacturing technologies as intended.

Knowledge-intensive services can therefore be considered from different perspectives. On the one hand, it is a certain type of complex services (engineering and technical; information, consulting, education, management ones) that have a new unique value for customers and that cannot be produced without the participation of highly professional experts possessing advanced knowledge of technology and without infrastructure, including research infrastructure, for constant generation and renewal of such knowledge. On the other hand, knowledge-intensive service can be defined as a specific segment of the service market that is characterized by service processes' embeddedness in production. This determines the heterogeneous composition of market players and constant involvement of both the provider and the customer in the process of service delivery.

The "watershed" between traditional and knowledge-intensive services lies in:

- the way the service is provided. Traditional services are delivered using the same means, instruments and personnel regardless of the service subject. Knowledge-intensive services call for a constant search for new methods, the development of customized solutions, team rotations, and external intellectual and machine capital;
- commercial value. Traditional services ensure the comfortable and efficient running of the main production process, product or system without any considerable changes to their key components. Knowledge-intensive services are highly innovative and bring profound changes to business models and the behavior of economic agents that interact in the process of service design and delivery;
- provision of analytical support for service delivery. The delivery of traditional and knowledge-intensive services lasts over long periods of time, and the quality of the service cannot be assessed until the delivery is complete. The use of analytical instruments, software and diagnostic programs for actual and remote monitoring can, however, enable the contractor to get a detailed snapshot of the phases of knowledge-intensive service production, to forecast potential risks and make appropriate adjustments.

3 CONCEPTUAL MODEL OF KNOWLEDGE-INTENSIVE SERVICE IN THE ENERGY SECTOR

The growing role of the service sector in energy has to do with a number of internal and external factors. On the one hand, global energy business is moving towards a structural model in which energy companies are increasingly focused on energy production management, asset management and projects of building new energy facilities or refurbishing existing ones. The majority of associated works aimed at technological search, forecasting market and customer behavior, engineering, and repair and maintenance of complex energy equipment, and logistics are executed by third party service companies.

Additionally, many energy companies cannot afford to have skilled employees and knowledge-intensive equipment for performing hi-tech works amid a transition to a new energy paradigm. This is particularly true when demand for such works is intermittent. Using the services provided by specialist service organizations enables energy businesses to shift the risk of employee and equipment underutilization to them. Such an approach enables service organizations to focus on performing selected complex technological operations and utilize the available personnel and their know-how to the full.

At the same time, the energy sector that is naturally prone to inertia is under pressure from an innovation avalanche triggered by Industry 4.0. This sets the background for three major trends that, according to experts from Bain&Company, determine the need for building knowledge-intensive service infrastructure in the energy sector: 1) electrification of an overwhelming majority of production processes in various industries; 2) decentralization as a form of independent market behavior of customers, and 3) digitalization that allows open communication between the grid operator and any of its elements (Table 2).

Industry-specific services are entering a new development stage; their scale is growing drastically. Knowledge-intensive services are becoming the key driver and responsibility center of progress in energy, focusing on the adaptation of new technologies, digital ones above all, to the desirable market structure. The authors could identify three key application areas for knowledge-intensive services in the modern energy sector (Table 3).

Here are some examples of digital services that have shown the best growth in the global energy market [30]–[32]:

- German power trader Next Kraftwerke, founded in 2009, is the operator of a 1-GW virtual power plant that pools together 2,400 installations (fueled by biogas, biomass, and even CHPPs). The company has an analytical platform that is capable of creating an operational schedule for the installations using real-time market data, including electricity prices and weather forecasts.

Table 2: Trends determining the need for building a knowledge-intensive service industry in the global energy sector *(adapted from [29])*.

Trend	Goal	Key technologies
Electrification	Critical for goals of carbon-free energy; sustainable development of the grid and regional energy sector	Electric vehicles, charging infrastructure integrated into grid; Electric heating; Electrothermics
Decentralization	Makes customers active elements of the energy market	Demand response; Distributed generation, accumulation and storage of energy; Microgrids
Digitalization	Allows for open, real-time communication and data exchange between the grid operator and the grid; automation	Smart cities (buildings, industrial facilties); Smart meters; Digital platforms for energy data aggregation; Internet of things

Table 3: Examples of activity areas for knowledge-intensive services in the energy sector in various segments of the value chain.

<table>
<tr><th>Energy Sector / KIS</th><th>Generation (incl. Small-scale)</th><th>Transmission</th><th>Distribution</th><th>Sale</th><th>Households and industrial energy facilities</th><th>Effects</th></tr>
<tr><td rowspan="5">Asset lifecycle management</td><td colspan="5">Remote monitoring of energy equipment parameters</td><td rowspan="5">• Longer lifespan of assets
• Productivity growth
• Optimization of management and technological solutions</td></tr>
<tr><td colspan="4">Strategy-aligned asset management</td><td></td></tr>
<tr><td colspan="3">Digital analytics and intelligent planning</td><td colspan="2" rowspan="3"></td></tr>
<tr><td colspan="3">Use of robots for mechanical works</td></tr>
<tr><td colspan="3">Smart grid reliability management</td></tr>
<tr><td rowspan="3">Energy market optimization and customer relations</td><td>Virtual power plants</td><td colspan="2">Electricity storage</td><td colspan="2">Control of clusters of energy consuming equipment bundled into a single network</td><td rowspan="3">• Demand aggregation
• Increased grid flexibility
• Loss reduction
• Real-time price management
• Hyper-personalization</td></tr>
<tr><td colspan="5">Platforms for demand response and real-time consumption control</td></tr>
<tr><td colspan="3"></td><td colspan="2">Smart metering</td></tr>
<tr><td rowspan="4">Development of organizational flexibility of energy market agents</td><td colspan="4">Preparation of teams and personal for technological modernization</td><td></td><td rowspan="4">• Emergency response to external threats and challenges
• Anticipatory competency development</td></tr>
<tr><td colspan="5">Corporate systems and mechanisms of knowledge transfer and advanced experience exchange</td></tr>
<tr><td colspan="5">Energy (incl. digital) consulting</td></tr>
<tr><td colspan="5">Research and development projects in partnership with universities</td></tr>
</table>

- Alliander (Netherlands) is currently piloting projects that match supply and demand at the local level in real time. Partnering with CGI, an IT solutions provider, Alliander is using its PowerMatcher software as the foundation for Realtime Energy eXchange (REX). This is expected to be rolled out to 800,000 streets across 14 municipalities. At a household level, REX enables nonessential appliances to be run at times when supply is abundant and power prices are low. Alliander and CGI are also researching potential ways for REX to take advantage of the growing increase in electric car ownership in the Netherlands, for example by using the batteries of electric vehicles as a flexible storage solution.
- PowerShop (Australia) is an integrated energy solution provider, offering consumers the ability to monitor and purchase electricity online, on mobile and via social media. From their smartphones, customers can monitor home energy consumption and choose the source of their electricity, such as from alternative energy projects or large power plants.
- US company General Electric has recently launched its flagship knowledge-intensive service, Predix, a platform for industrial-scale analytics, which connects machines, sensors, control systems and devices to capture and interpret data from manufacturing systems. Predix is a highly flexible product, with its application including monitoring and diagnostics for rail transportation, cloud imaging for healthcare, and nondestructive testing remote collaboration for oil and gas utilities. Predix is an example of a unique knowledge-intensive service platform that a big energy corporation was able to create through cooperation with partners from adjacent markets, thus adding advanced experience, knowledge and competencies to its own practice.

The traditional grid-based power industry has also seen the emergence of areas in which knowledge-intensive services are of critical importance. This, above all, refers to new principles of asset management and repair and maintenance arrangements for power plants and transmission networks (boilers, turbines, generators, power transformers, transmission lines, switchgear, breakers, nuclear power reactors, penstocks etc.) The model of knowledge-intensive services for repair and management builds upon a strategy of risk and performance management that envisages ranking equipment by reliability, analyzing the state of a component and the piece of machinery as a whole, assessing the risk of failure and potential damage from the economic and environmental points of view and technological implications.

Technologically, the process could be described as follows: RFID tags are attached to energy equipment. The tags capture machine performance data that can be read by shop floor workers using tablet PCs or smartphones or can be transferred to a data storage hub in real-time mode. The data is then analyzed using a Big Data technology and an equipment failure forecast is generated in real-time mode, making it possible to timely fix malfunctioning units, thus reducing potential losses from downtime. In order to switch over to the model, energy companies have to use modern diagnostics tools and predictive analytics.

A conceptual model of knowledge-intensive services in the energy sector is depicted in Fig. 2 The goal is to demonstrate the interconnectedness of various types of knowledge-intensive services and to reflect the scale of the service industry. The cube has been selected as a visual metaphor for the model. The cube is split into layers along the X axis, indicating the corporate, regional and industrial service segments. Each axis is divided into three, representing the complexity of delivering certain knowledge-intensive services. The division also makes it easier to classify individual groups of services within a coordinate system.

A comprehensive implementation of the model is an extremely laborious task. In fact, it means building large energy service infrastructure based on systemic principles. Such infrastructure should be synchronized with technology and set the "rules of the game" for different energy market stakeholders, who are used to acting in line with their own business, legal and financial laws. Some countries, e.g. the US, Great Britain, Germany, and Canada, have been successfully embracing the idea of the system configuration of the market for knowledge-intensive services. This usually means creating technology interfaces that are universal for all market participants, a corresponding innovative and regulatory environment, corporate and municipal support tools for individual projects. That brings closer the interests of energy companies, contracts and consumers to a certain degree.

The author is convinced that the required infrastructure should be built within the dimensions of the presented conceptual model that indicate several closely connected areas.

1. As regards energy market optimization.

- Encouraging competition between energy market participants – especially energy service companies and energy distributors – operating in the segments of the value chain that are the closest to the consumer.
- Ensuring access of various categories of consumers (for example, owners of electric vehicles) to the digital infrastructure of the market in order to enable vehicle-to-grid operations.
- Incentives (in the form of legal and financial mechanisms) for technology entrepreneurship that is viewed as the driver of the industry-specific innovative process.
- A more flexible tariff policy and legalization of dynamic pricing in order to enable the implementation of demand side and consumption management programs.

As part of this set of measures, one needs to highlight legal provisions for the mass adoption of the internet of things, and the processing of consumers' energy data [29], [33]–[36]. All too often, outside actors of the telecommunications market who are capable of providing such services have limited or no access to the required information. This has to do with a risk of cyber security infringements and data breaches, which could much more serious implications in the energy sector as compared to other industries due to the need to ensure the reliability and stability of power supply. At the same time, this barrier impedes the development of service companies that make it possible to monitor and control the market in which the number of participants – both energy producers and consumers – has been growing drastically.

2. As regards asset lifecycle management.

- Creation of joint platforms of energy companies, energy equipment producers and telecom businesses for the sake of a transition to advanced forms of repair and maintenance and intelligent reliability management.
- Development of a universal system of communication between household energy networks and top-level dispatch control centers that is based on unified data models.
- Adoption of forward-looking technical standards that harmonize the development of various elements of the grid – from low-voltage networks to operators in all sectors, from power plants to end consumers, from markets to technological processes.

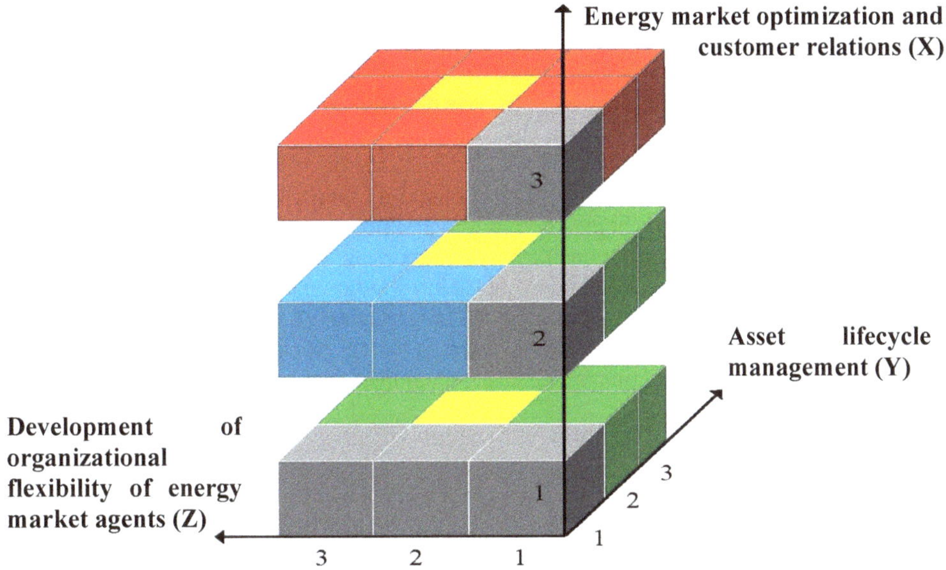

Grey Zone – IT-services

(1, 1, 1)
(2, 1, 1)
(3, 1, 1) Services supporting business processes and customers' interaction with energy markets (applications, software, data aggregation solutions, data bases)

(1, 1, 2) In-house consulting and R&D-services

(1, 1, 3) Development of organizational flexibility of energy market agents

Yellow Zone – services for personnel training and competency development

Green Zone – services supporting the adoption of new principles of repair and maintenance of energy equipment, revamping and technical upgrading of energy facilities

(1, 2, 1) Use of robots for individual operations and mechanical works

(1, 3, 1) Remote monitoring and energy equipment performance analysis

(1, 2, 3) Digital twins

(1, 3, 2) Optimization of manufacturing logistics

(1, 3, 3) Digital analytics and intelligent planning

(2, 2, 1) Servicing of smart devices and charging infrastructure, incl. in transportation

(2, 3, 1) Energy service contracts for energy conservation

(2, 3, 2) General engineering using EPCM and BOOT schemes

Blue Zone – demand side management services

(2, 1, 2) Smart metering

(2, 1, 3) Electricity storage

(2, 2, 3) Virtual power plants

(2, 3, 3) Real-time demand response and consumption management platforms

Red Zone – “institutional” services

(3, 1, 2) Digital financial services

(3, 2, 1) Experimental platforms for testing new technologies

(3, 1, 3) Cyber security and energy data storage

(3, 3, 1) Digital consulting

(3, 2, 3) Industry and cross-industry knowledge transfer, including monitoring mechanisms for research advances

(3, 3, 2) Intelligent reliability management

(3, 3, 3) Analytical platforms for balancing supply and demand

Figure 2: A conceptual model of knowledge-intensive service in the energy sector.

- A transition to state regulation models that encourage the creation of soft assets by energy companies rather than hard assets investments.

The latter step might seem controversial, but it has to be noted that digitalization in and of itself is an extremely costly process (just imagine the cost (including downtime cost) of fitting the boilers/turbines of power plants with digital sensors or building a chain of charging stations for electric vehicles in a city) that is bound up with creating a supporting service infrastructure, which entails high fixed and variable costs. That's why many energy companies continue to build their business on maximizing the sales of their main product – electricity and thermal energy. For some types of knowledge-intensive services, such as EPC, EPCM, and BOOT schemes in engineering and construction of energy facilities, this barrier does not play the key role, but it does when it comes to implementing demand side management programs in a region or executing a complex energy service contract.

3. As regards greater organizational flexibility of energy market agents.

- Designing a methodology for integrated management of new grid facilities: virtual power plants, digital substations, flexible power distribution systems, energy storage systems, power supply networks for smart buildings, intelligent electric transport infrastructure.
- Adoption by energy companies and industry organizations of proactive management based on the early detection of opportunities and threats, methods of predictive analytics, principles of systems engineering and agile methods.
- Building close cooperation between energy companies, universities and science for conducting breakthrough research, knowledge generation and transfer, construction of new experimental platforms for performing tests, and commercializing innovations and bringing them to the market.
- Creation of a system of anticipatory training of personnel and teams for the technology modernization of the energy sector.

The latter issue is unfortunately often overlooked in both theory and practice. The author is convinced that without having a system that makes it possible to predict the competencies that will soon be required and to nurture them in personnel it is pointless to speak about moving on to a new energy paradigm and to knowledge-intensive service at its core in particular.

The experience of a team of experts from Ural Federal University, including the author, shows that the problem can be solved by creating corporate platforms, with initiative coming from energy companies in cooperation with educational and scientific organizations.

The pilot testing of the projects “Technological Leadership Incubator” and “Training for Breakthrough Teams” on the premises of several Russian energy companies confirms the idea that the infrastructure of knowledge-intensive energy services must include centers of excellence that have the required scientific and technical background and ensure a transition from R&D results to services with competitive and replicable properties and perform the verification of the declared properties of new manufacturing technologies for potential customers.

4 CONCLUSION

The present study is aimed at giving a comprehensive picture of the complexity and multifacetedness of knowledge-intensive services in the energy sector that is experiencing an inflow of digital technologies in production, the growth of distributed generation and micro-grids, the emergence of numerous cogeneration energy facilities that require functional and technological compatibility, and new markets, for example, for energy storage devices. At the same time, consumers are getting increasingly proactive and the sector itself is integrating with other critical infrastructure industries.

The problem of knowledge-intensive services in the energy sector has found little coverage in academic literature and is at an early stage of its development in practice. Consequently, the list of identified barriers and recommendations is by no means exhaustive. Additional theoretical and empirical analysis is required to reveal the interdisciplinary nature of the subject in a more profound way and to verify the parameters of the proposed conceptual model of knowledge-intensive services. The author views as promising and intends to work on the following areas.

4. Methodology of developing flexible manufacturing and organizational systems ensuring pre-emptive response to external challenges and threats in the energy sector.
5. Market platforms and eco-systems in the energy sector as the implementation of new forms of relationships between producers and consumers of energy services.
6. Cross-industry incentive mechanisms for competition in the market of knowledge-intensive energy services.
7. Systems of knowledge transfer in the “energy business – universities” chain to facilitate anticipatory development of competencies and training of skilled personnel.

ACKNOWLEDGEMENT

The work was supported by Act 211 of the Government of the Russian Federation, contract № 02.A03.21.0006.

REFERENCES

[1] Growth in Services, Fostering Employment, Productivity and Innovation, OECD, 2005. www.oecd.org/economy/34749412.pdf. Accessed on: 20 Aug. 2018.

[2] Nählinder, J., Innovation and Employment in Services. The case of Knowledge Intensive Business Services in Sweden, 2005. www.diva-portal.org/smash/get/diva2:20573/FULLTEXT01.pdf. Accessed on: 20 Aug. 2018.

[3] Sultan, M., The Tertiary Sector Is Going to Dominate the World Economy; Should We Worry? Online. https://mpra.ub.uni-muenchen.de/14681/1/MPRA_paper_14681.pdf. Accessed on: 20 Aug. 2018.

[4] Buera, F.J. & Kaboski, J.P., The Rise of the Service Economy. *American Economic Review*, **102(6)**, pp. 2540–69, 2012.

[5] Preissl, B., Service sector dynamics and service regulation, 2000. Online. www.tips.org.za/research-archive/trade-and-industry/trade-growth-dynamics/item/92-service-sector-dynamics-and-service-regulation. Accessed on: 20 Aug. 2018.

[6] Johnston, A. & Huggins, R., Regional growth dynamics in the service sector: The determinants of employment change in UK regions, 1971–2005. *Growth and Change*, **49**(1), pp. 71–96, 2018. DOI: 10.1111/grow.12221.

[7] Gölpek, P., Service sector and technological developments. *Procedia—Social and Behavioral Sciences*, **181**, pp. 125–130, 2015. DOI: 10.1016/j.sbspro.2015.04.873.

[8] Preissl, B., Research and technology organisations in the service economy. developing analytical tools for changing innovation patterns. *Innovation. The European Journal of Social Science Research*, **19(1)**, pp. 131–146, 2006. DOI: 10.1080/13511610600608039.
[9] Pino, G., Capestro, M., Guido, G., Tomacelli, C. & Abate, M., Knowledge-intensive services and local development: An empirical analysis of networks, channels and customization processes. *Local Economy*, **31(3)**, pp. 359–376, 2016. DOI: 10.1177/0269094216642750.
[10] Shi, X., Wu, Y. & Zhao, D., Knowledge intensive business services and their impact on innovation in China. *Service Business*, **8(4)**, pp. 479–498, 2014. DOI: 10.1007/s11628-013-0202-1.
[11] Fisher, A.G.B., Primary, secondary and tertiary production. *Economic Record*, **15**, pp. 24–38, 1939.
[12] Clark, C., *The Conditions of Economic Progress*, Macmillan: London, p. 504, 1940.
[13] Martinez-Fernandez, M.C., Soosay, C., Krishna, V.V., Turpin, T. & Bjorkli, M., Knowledge-Intensive Service Activities (KISA) in Innovation of the Software Industry in Australia, 2005. Online. www.uws.edu.au/__data/assets/pdf_file/0014/33431/KISA_-_Innovation_of_the_Software_Industy_in_Australia.pdf. Accessed on: 20 Aug. 2018.
[14] Knowledge-Based Industries (KBI). Online. www23.statcan.gc.ca/imdb-bmdi/document/2514_D2_T9_V1-eng.pdf. Accessed on: 20 Aug. 2018.
[15] Miozzo, M. & Grimshaw, D., Modularity and innovation in knowledge-intensive business services: IT outsourcing in Germany and the UK. *Research Policy*, **34(9)**, pp. 1419–1439, 2005. DOI: 10.1016/j.respol.2005.06.005.
[16] Hyppönen, H., Saalasti-Koskinen, U., Perala, M-L. & Saarikalle, K., Knowledge-Intensive Service Activities (KISA) in Health and Social Care Innovation Process: Towards Seamless care for older people in Kuopio Home Care. Online. www.oecd.org/sti/inno/34745295.pdf. Accessed on: 20 Aug. 2018.
[17] Kryukov, V.A. & Tokarev, A.N., Features of spatial development of oil and gas service: global trends and lessons for Russia, [in Russian]. *Region: Economics and Sociology*, **3**, 189–214, 2017. DOI: 10.15372/REG20170310.
[18] Miles, I. & Boden, M., *Are Services Special? Services and the Knowledge-based Economy*, Routledge: London, p. 285, 2000.
[19] Kox, H. & Rubalcaba, L., Business services and the changing structure of European economic growth. *CPB Memorandum*, **183**, 2007, Online. https://mpra.ub.uni-muenchen.de/3750/1/MPRA_paper_3750.pdf. Accessed on: 20 Aug. 2018.
[20] Eichengreen, B. & Gupta, P., The two waves of service sector growth. *Oxford Economic Papers*, **65(1)**, pp. 96–123, 2013.
[21] Homburg, C., Workman, J.P. & Jensen, O., Fundamental changes in marketing organization: The movement toward a customer-focused organizational structure. *Journal of the Academy of Marketing Science*, **28**(4), pp. 459–478, 2000. DOI: 10.1177/0092070300284001.
[22] Wong, P.-K. & He, Z.-L., A comparative study of innovation behaviour in singapore's kibs and manufacturing firms. *The Service Industries Journal*, **25**(1), pp. 23–42, 2005. DOI: 10.1080/0264206042000302397.
[23] Muller, E. & Doloreux, D., The key dimensions of knowledge-intensive business services (KIBS) analysis: a decade of evolution, Fraunhofer Institute for Systems and Innovation Research (ISI), 2007. Online. www.econstor.eu/obitstream/10419/29335/1/610017543.pdf. Accessed on: 20 Aug. 2018.

[24] Allmendinger, G. & Lombreglia, R., Four strategies for the age of smart services. *Harvard Business Review*, **83**, pp. 131–145, 2005.

[25] Wunderlich, N.V., von Wangenheim, F. & Bitner, M.J., High tech and high touch: a framework for understanding user attitudes and behaviors related to smart interactive services. *Journal of Service Research*, **11**, pp. 1–18, 2012.

[26] Vuori, E.K., Knowledge-intensive service organizations as agents in a business ecosystem. *Proceedings of ICSSSM '05*, **2**, pp. 908–912, 2005. DOI: 10.1109/ICSSSM.2005.1500123.

[27] The New Zealand Sectors Report 2014, Knowledge-Intensive Services. Online. www.mbie.govt.nz/info-services/sectors-industries/sectors-reports-series/pdf-image-library/knowledge-intensive-services-report/knowledge-intensive-services-report.pdf. Accessed on: 20 Aug. 2018.

[28] Rzevski, G., Complexity as the defining feature of the 21st century. *International Journal of Design & Nature and Ecodynamics*, **10**(3), pp. 191–198, 2015. DOI: 10.2495/DNE-V10-N3-191-198.

[29] The Future of Electricity, New Technologies Transforming the Grid Edge. Online. www3.weforum.org/docs/WEF_Future_of_Electricity_2017.pdf. Accessed on: 20 Aug. 2018.

[30] Digital Transformation of Industries, Electricity Industry. White Paper, 2016. Online. www.accenture.com/t20170411T120540Z__w__/us-en/_acnmedia/Accenture/Conversion-Assets/WEF/PDF/Accenture-Electricity-Industry.pdf. Accessed on: 20 Aug. 2018.

[31] Bessa, R.J. & Matos, M.A., The role of an aggregator agent for EV in the electricity market. *7th Mediterranean Conference and Exhibition on Power Generation, Transmission, Distribution and Energy Conversion (MedPower 2010)*, pp. 1–9, 2010. DOI: 10.1049/cp.2010.0866.

[32] Weber, G., Mateescu, R.M., Lange, S. & Rauch, M., Knowledge intensive business services (KIBS) in the context of changing energy economics in Germany. *Amfiteatru Economic*, **18**(41), pp. 89–103, 2016.

[33] Quéré, S., Delmas-Patterson, G. & Berezin, A., A new paradigm of governance for a carbon-pricing system. *International Journal of Energy Production and Management*, **1**(2), pp. 192–205, 2016. DOI: 10.2495/EQ-V1-N2-192-205.

[34] Digital Economy: Global Trends and Practice of Russian business, [in Russian]. Analytical Report, National Research University "Higher School of Economics." Online. https://imi.hse.ru/pr2017_1. Accessed on: 20 Aug. 2018.

[35] Annunziata, M. & Bell, G., Powering the Future. Leading the digital transformation of the power industry, 2015. Online. www.res4med.org/wp-content/uploads/2017/05/digital-energy-transformation-whitepaper.pdf. Accessed on: 20 Aug. 2018.

[36] Booth, A., Mohr, N. & Peters, P., The digital utility: New opportunities and challenges, 2016. Online. www.mckinsey.com/industries/electric-power-and-natural-gas/our-insights/the-digital-utility-new-opportunities-and-challenges. Accessed on: 20 Aug. 2018.

DIAGNOSTICS OF COMPETITIVENESS OF POWER-GENERATING COMPANIES

ALEXEY DOMNIKOV, EKATERINA ANTIPOVA & LIUDMILA DOMNIKOVA
Academic Department of Banking and Investment Management, Ural Federal University Named after the First President of Russia B.N. Yeltsin, Russia

ABSTRACT

The paper presents the results of a study of the processes of development of power-generating companies in the conditions of high risks and uncertainty. To reduce the impact of uncertainty on the prospects of energy business development in the field of energy generation, a system of diagnostics of the competitiveness of power-generating companies is proposed. This system makes it possible to identify the threats associated with the reduction of competitiveness and provide information and analytical support to the management of companies to create effective management decisions. The developed system of diagnostics is based on mathematical modeling using procedures of discriminant and cluster analysis. The diagnostic process identifies and evaluates the factors that affect the increase in risks and reduce the investment attractiveness of the energy business in the energy generation sector by centralized energy sources. This makes it possible to use the potential advantages of the market situation to improve the competitiveness of power-generating companies. The system of diagnostics of competitiveness of the power generating companies developed by authors is the effective analytical tool which raises quality and objectivity of the made administrative decisions. This will make it possible to avoid the appearance of adverse consequences, which may lead to a decrease in the competitiveness of power-generating companies in the electricity and heat markets.

Keywords: power industry, efficiency, competition, strategy, reliability, risks, mathematical economic models, uncertainty, centralized energy sources.

1 INTRODUCTION

Increased competition in Russian power industry caused by recent reforms in this sector brought into existence two interconnected systems in the field of power generation. The first system includes the power-generating facilities of large nuclear, hydraulic, and thermal power plants which are mainly used for maintaining the power balance in electricity supply networks, ensuring the safe operation of the state unified energy supply system, and for maintaining the standard power quality in this system. The second power generation system is located at the territorial level. It forms a so-called territorial power generation system that has a separate niche in the territorial energy market.

The development of the territorial power generation system depends on its internal organization on the one hand, and on its interaction with the environment and climatic conditions on the other hand. From the system approach point of view, it possesses the integral unity of interconnected parts, i.e. the centralized and the distributed energy generation systems each of which can be divided into cogeneration power sources and separate cogeneration plants that can be characterized by the lower level of concentration and centralization of power-generating facilities.

The centralized energy cogeneration system includes thermal-type cogeneration facilities, the heat and power plants, which constitute the basis of a centralized heat supply system. These facilities cover 37% of the installed capacity of all power stations and up to the half of the thermal power generation in Russia. Usually these facilities are owned by territorial power generation companies. Steam turbine plans form the basis of the generating capacities

WIT Transactions on Ecology and the Environment, Vol 222,
www.witpress.com, ISSN 1743-3541 (on-line)
doi:10.2495/EQ180031

in the centralized cogeneration systems (up to 80%), however combined cycle gas turbine plants and gas turbine plants are considered to be very competitive.

The distributed energy cogeneration system consists of high-capacity and medium-capacity cogeneration power facilities located near power demand centres, including facilities catering the needs of separate consumer groups and units (typically located in remote areas of a region).

Therefore, the competitiveness of power generation companies can be described as a system of relations between the thermal and electrical power generation companies as regards creating, implementing, and maintaining competitive advantages. These organizational and economical features influence the efficiency of cogeneration power facilities and allow the improvement of cost and quality parameters of supplied energy.

For the evaluation of competitive advantages of cogeneration and conditions of the territorial energy market positioning, it was necessary to create a set of procedural tools allowing the user to determine the level of competitive capacity of a power generation company.

This aim can be achieved using a power-generating company competitiveness diagnostics method, created with an aim to:

- identify the causes of development risk occurrence and increase;
- identify the measures and capabilities of development risk minimization;
- develop the information and methodological basis for determining the perspective competitiveness level.

The identified goals can be achieved only by means of detailed analysis of power-generating company development tendencies using system analysis methods. Besides that, it is possible to use regular methods with heuristic procedures based on expert analysis, since the diagnostics of competitiveness of a power-generating company is similar in nature to a simulation system [1]–[3].

2 PROCEDURAL ASPECTS OF POWER-GENERATION COMPANY COMPETITIVENESS DIAGNOSTICS

From the procedural point of view, the determination of limit levels of indicators is of a fundamental importance for diagnosing the competitiveness of a power generation company. Too high or two low values of these indicators may cause the formation of negative processes. The variation of these indicators may have an individual effect or may be combined with other factors increasing negative consequences that may affect the competitiveness level of a power-generating company [4], [5].

Proposed competitiveness diagnostic diagram for power-generating companies is shown in Fig. 1 below.

When conducting the diagnostics of power-generating company competitiveness it is very important to take into account the range of indicators reflecting the condition of the territorial energy market and competitive capabilities of each energy source. This approach allows identifying competitiveness level reduction causes and concentrating resources on the weakest units of a power-generating company [1], [6].

The first stage of the diagnostics consists of the preparation of time series of indicators (this is achieved by means of monitoring the business environment of the territorial energy market). For this purpose, the developed database is associated with an indicator analysis algorithm. It allows increasing the accuracy of diagnostics and processing larger amounts of data for every monitoring object for the given time period. Then, the total quantity of obtained indicators is divided into indicator blocks corresponding to the monitoring objects of the territorial energy market business environment.

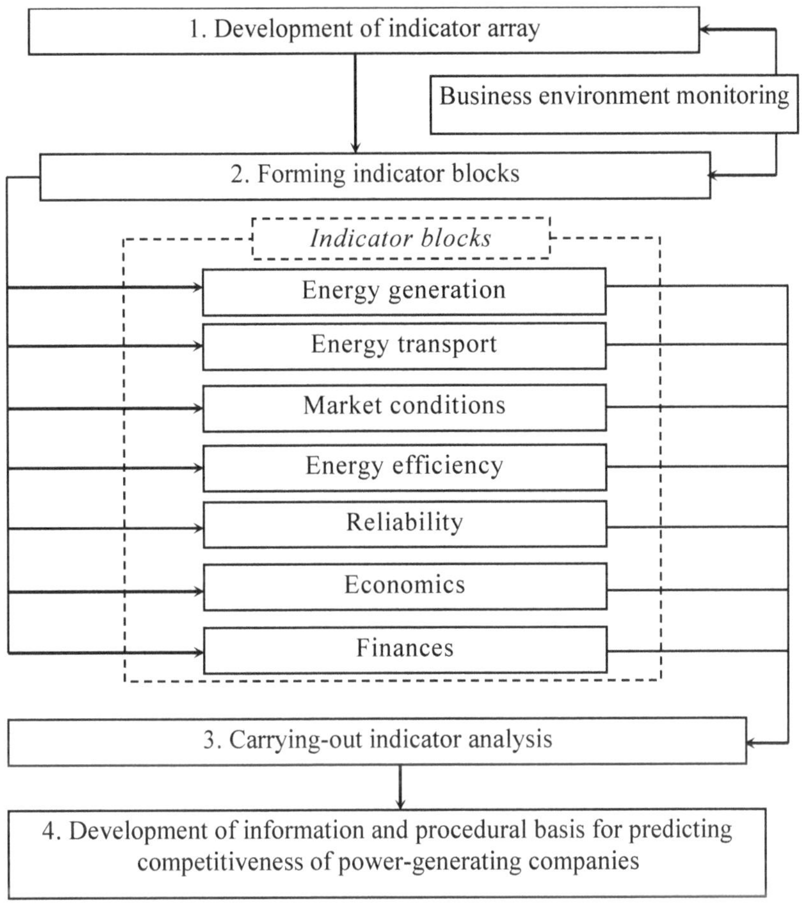

Figure 1: Diagram of procedure for diagnosing competitiveness of power-generating companies.

These indicator blocks are used for the competitiveness diagnostics of power-generating companies considering the distinctive features of each indicator included in a block.

The indicator blocks developed for the competitiveness diagnostics of a power-generating company shall comply with the following principles:

- they shall reflect the regularities of competitive development of power cogeneration systems;
- it is necessary to identify the most significant monitoring objects that can be classified as criterion-type objects which can be used for assessing the competitiveness level;
- monitoring objects shall be aggregated according to similar features of their nature and the character of influence on the development process;
- information support of parameters determining the value of each analyzed indicator;
- convenience and possibility of classification of the competitiveness levels by the monitoring objects in qualitative and quantitative forms.

- At the second stage the indicators are aggregated into indicator blocks. It is necessary for considering the peculiar features if realization of competitive advantages of a power-generating company.
- At the third stage an indicator analysis is carried out for each indicator block and for the whole object basing on cluster and discriminative analysis principles [7]–[13].
- At the final fourth stage the information and procedural basis for the prediction of competitiveness level of a power-generating company is prepared.

3 ASSESSMENT OF COMPETITIVENESS OF POWER-GENERATING COMPANY BUSINESS UNITS

The practical implementation of the proposed approach to the indicator analysis was tested by means of the competitiveness diagnostics carried-out for the power stations of Sverdlovskiy branch of T+Group JSC company (installed electrical power: 1256 MW, installed thermal power: 5886 Gcal/h). Sverdlovskiy branch of T Plus Company Group includes power generation and heat distribution facilities located in seven cities of Sverdlovsk region. The branch includes 6 heat and power plants: Nizhneturinskaya, Akademicheskaya, Verkhneturinskaya, Novo-Sverdlovskaya, Sverdlovskaya, and Pervouralskaya.

Series of imitation calculations were conducted basing on statistical data provided in the company reports for the year 2017. The first stage of analysis aimed at determining the level of competitiveness of the power-generating company business-units involved the calculation of normalized threshold values of competitively levels for each indicator block (Table 1). Upon that, these values were used as a basis for the evaluation of competitiveness of the power-generating company (Sverdlovsk branch).

The authors' experience in the field of diagnostics tells that the division into three groups is not sufficient for differentiating the qualitative status of an object and does not allow the reliable determination of the rate of reaction to the decrease in the competitiveness level for each business-unit of the power-generating company. Therefore, it is reasonable to identify three levels consisting of different stages of reduction of the competitiveness level in the transient and critical groups (Table 2).

Table 1: Normalized threshold values.

Indicator block	Threshold level					
	B	*C*	*D*	*E*	*F*	*G*
Energy generation	0.258	0.371	0.484	0.597	0.731	0.866
Energy transport	0.198	0.298	0.398	0.498	0.665	0.833
Market conditions	0.456	0.556	0.656	0.756	0.837	0.919
Energy efficiency	0.245	0.325	0.405	0.485	0.657	0.828
Reliability	0.387	0.524	0.661	0.798	0.865	0.933
Economics	0.309	0.415	0.521	0.627	0.751	0.876
Finances	0.248	0.339	0.431	0.523	0.682	0.841
Overall situation	0.300	0.404	0.508	0.612	0.741	0.870

Table 2: Classification of statuses as per the level of competitiveness.

Groups	Level of competitiveness (LoC)	Designation
Normal (N)	Weak	*A*
Transient (T)	Low	*B*
	Medium	*C*
	Unstable	*D*
Critical (K)	High	*E*
	Ominous	*F*
	Grave	*G*

Table 3: Results of power-generating company business units competitiveness diagnostics.

Business unit (heat and power plant)		Indicator block							Overall situation	Rating
		Energy generation	Energy transport	Market conditions	Energy efficiency	Reliability	Economics	Finances		
Pervouralskaya	*r*	0.326	0.308	0.487	0.472	0.529	0.418	0.447	0.497	3
	h	*C*	*D*	*C*	*E*	*D*	*D*	*E*	*D*	
Akademicheskaya	*r*	0.154	0.134	0.259	0.248	0.388	0.312	0.247	0.290	1
	h	*B*	*B*	*B*	*C*	*C*	*C*	*C*	*B*	
Nizhneturinskaya	*r*	0.484	0.614	0.706	0.397	0.809	0.621	0.798	0.738	6
	h	*D*	*F*	*E*	*D*	*F*	*E*	*G*	*F*	
Sverdlovskaya	*r*	0.254	0.322	0.543	0.395	0.511	0.589	0.423	0.506	4
	h	*B*	*D*	*C*	*D*	*C*	*E*	*D*	*D*	
Verkhneturinskaya	*r*	0.325	0.196	0.612	0.307	0.651	0.573	0.487	0.525	5
	h	*C*	*B*	*D*	*C*	*D*	*E*	*E*	*E*	
Novo-Sverdlovskaya	*r*	0.198	0.271	0.496	0.438	0.315	0.419	0.257	0.399	2
	h	*B*	*C*	*C*	*E*	*B*	*D*	*C*	*C*	

Note: *r* – calculated value; *h* – current status.

The results of the power-generating company business unit competitiveness evaluation (as exemplified by Sverdlovsk branch of T+ Group JSC) carried-out using the indicator blocks are given in the Table 3 with overall situation assessment index and ratings.

The results of the diagnostics showed that Akademicheskaya and Novo-Sverdlovskaya heat and power plants have the highest competitiveness level, and Nizhneturinskaya and Verhneturinskaya heat and power plants have the lowest competitiveness level.

In the course of diagnostics, it had been established that the most significant risk factor related to the decrease in the competitiveness of the power-generating company is the insufficient reliability of heat distribution networks that require regular repairs. This is the main reason why many heat power consumers reject centralized heat supply and opt to use individual power sources instead. Power supply network losses are influenced by a power generation equipment reliability factor that impacts on the index of availability of a power source for serving a load. However, one of the main causes of power network loss is the deterioration of network equipment due to high wear. Besides that, the increase in the share of low and medium voltage consumers, that is currently observed, causes power network losses and changes in the electrical load pattern, thus affecting the efficiency of cogeneration equipment application in heat and power plants. Probably this situation can also be attributed to insufficient investments that cause the decrease in the reliability of cogeneration plants due to high wearing.

The development of a decentralized heat supply system leads to the increase in competition and decrease in the share of a power-generating company in the thermal energy market. This, in turn, affects the technical and economical performance of cogeneration plants and also causes higher aggregate fuel consumption and escalation of costs.

The increase in the installed power of heat and power plants is a very significant competitiveness growth factor that in turn requires additional investments. It is worth mentioning, that the reduction of the competitiveness level is largely influenced by the operating conditions of cogeneration plants that are often far from optimal. Also, the important point is the fact that fuel management optimization and used fuel type diversification (where necessary) are often neglected.

It is evident that the one of the major impediments to the growth of competitiveness of the power-generating company business units is related to their financial improvement problems mainly caused by heavy indebtedness of power consumers. This factor significantly influences the investment potential and chosen financing options of a power-generating company.

4 CONCLUSIONS

The study of the global issue of competitiveness improvement in the electric power industry sector necessitated the development of a system for diagnosing the competitiveness of power-generating companies with a purpose to identify risks and assess their influence on territorial energy market positioning prospects. Analytical tools capable of identifying individual and integral characteristics of competitiveness were developed to achieve this aim.

In the future it is planned to use the proposed system for assessing the costs of development risks for the power-generating companies. It will allow analyzing the possibilities of realization of competitive advantages of cogeneration power sources. The proposed diagnostics system makes it possible to monitor changes in the business environment and take corrective actions that would allow the management of a power-generating company to timely minimize financial losses and use potential advantages of the current market situation.

It shall be noted that the significant enhancement of competitiveness of a power-generating company can be achieved by means of comprehensive solutions in the field of strategic goals prioritization: 1) increasing the efficiency of structure of generating plants; 2) fuel consumption optimization; 3) construction of highly flexible combined cycle gas turbine plants and cogeneration gas turbine plants. Proposed strategic priorities may be considered as a basis for increasing the competitiveness of a power-generating company at the local energy market and enhancing the economic efficiency of the electric power business.

ACKNOWLEDGEMENT

The work was supported by Act 211 Government of the Russian Federation, contract No 02. A03.21.0006 and Russian Foundation for Basic Research (RFBR), contract No 16-06-00317.

REFERENCES

[1] Domnikov, A., Khodorovsky, M. & Khomenko, P., Optimization of finances into regional energy. *Economy of Region*, **2**, pp. 248–254, 2014. DOI: 10.17059/2014-2-24.

[2] Domnikov, A., Chebotareva, G. & Khodorovsky, M., Unbiased investment risk assessment for energy generating companies: Rating approach. *International Journal of Sustainable Development and Planning*, **12**(7), pp. 1168–1177, 2017. DOI: 10.2495/sdp-v12-n7-1168-1177.

[3] Domnikov, A., Khomenko, P. & Khodorovsky, M., Value-based approach to managing the risks of investing in oil and gas business. *International Journal of Sustainable Development and Planning*, **12**(6), pp. 1085–1095, 2017. DOI: 10.2495/sdp-v12-n6-1085-1095.

[4] Domnikov, A., Chebotareva, G., Khomenko, P. & Khodorovsky, M., Risk-oriented investment in management of oil and gas company value. *International Journal of Sustainable Development and Planning*, **12**(5), pp. 946–955, 2017. DOI: 10.2495/sdp-v12-n5-946-955.

[5] Domnikov, A., Khomenko, P., Khodorovskiy, M. & Vlasov, V., Economic capital assessment method improvement based modifcation the exposure at default calculating method. *Audit and Financial Analysis*, **4**, pp. 113–117, 2013.

[6] Domnikov, A., Chebotareva, G. & Khomenko, P., Risk and profitability optimization of investments in the oil and gas industry. *International Journal of Energy Production and Management*, **2**(3), pp. 263–276, 2017. DOI: 10.2495/eq-v2-n3-263-276.

[7] Hayek, F., *The Meaning of Competition in Austrian Economics*, Hillsdale, pp. 264–280, 1991.

[8] Lance, G.N., A general theory of classificatory sorting strategies. 1. Hierarchical systems. *The Computer Journal*, **9**(4), pp. 373–380, 1967.
DOI: 10.1093/comjnl/9.4.373.

[9] Ward, J.H. Jr., Hierarchical grouping to optimize an objective function. *Journal of the American Statistical Association*, **58**(301), pp. 236–244, 1963.
DOI: 10.1080/01621459.1963.10500845.

[10] Schroeck, G. (ed.), *Risk Management and Value Creation in Financial Institutions*, John Wiley & Sons, Business & Economics: England, pp. 47–85, 2002.

[11] Jardine, N. & Sibson, R., *Mathematical Taxonomy*, London: John Wiley and Sons, p. 286, 1971.

[12] Hartigan, J.A., *Clustering Algorithms*, New York: Wiley, p. 351, 1975.

[13] Gordon, A.D., Identifying genuine clusters in a classification. *Computational Statistics & Data Analysis*, **18**(5), pp. 561–581, 1994. DOI: 10.1016/0167-9473(94)90085-x.

COVENANT OF MAYORS COMMITTED TO LOCAL SUSTAINABLE ENERGY: THE ITALIAN CASE

LUCA ADAMI, GIACOMO CASTAGNA, MARCO RAGAZZI & MARCO TUBINO
Department of Civil, Environmental and Mechanical Engineering, University of Trento, Italy

ABSTRACT

The Paris agreement fixed worldwide targets on the temperature increase below 2°C in the XXI century. Presently 176 countries ratified this framework. With the 20-20-20 package, the European Commission fixes three targets: (i) reduction of the European greenhouse gas emissions at least 20% below 1990 levels by 2020, (ii) 20% of EU energy consumption must come from renewable resources by 2020 and (iii) 20% reduction in primary energy use compared with projected levels, by improving energy efficiency. To achieve these global targets, Covenant of Mayor was launched in 2008 as a voluntary initiative for local administrations that decide to adopt a plan of interventions concerning greenhouse gas mitigations achieving and exceeding the EU climate and energy targets. In the present paper, we analyse the global results of the initiative during its first decade of activity, focusing on the Italian case at national level, since it is presently the leading country in terms of population involved.

Keywords: Covenant of Mayors, climate change, energy efficiency, local energy plans.

1 INTRODUCTION

Global studies on climate change show that the increased celerity of variations registered in the XX century [1], [2] can sensibly affect natural ecosystems [3], [4] as well as human evolution [5], with potentially drastic scenarios in the long term [6], [7]. This suggests that strategies for sustainable management of such ecosystems should focus on maintaining their resilience [8].

On this direction, the Paris agreement on climate can be considered the leading global framework on the topic, with the following aims [9]–[11]:

- to keep the global temperature rise in XXI century below 2°C above pre-industrial levels, possibly limiting this increasing trend to 1.5°C
- to strengthen the ability of countries to deal with the impacts of climate change.

The European Strategy for Climate change fixes three targets to make Europe more climate-resilient [12]–[15]:

(i) promoting action by Member States
(ii) promoting 'Climate-proofing' actions at European Union (EU) level
(iii) better informing decision-makers

Its low-carbon economy roadmap suggests that by 2050, EU should cut greenhouse gas emissions to 80% below 1990 levels and intermediate milestones to achieve this goal are 40% emissions cuts by 2030 and 60% by 2040 [16]. The first deadline is set in 2020, in which the goal is to reduce greenhouse gas (GHG) emission by 20% [17]. To reach this goal EU encourages several actions including all territorial levels, with local emphasis [18]–[21].

Under this perspective, the EU Covenant of Mayors (CoM) for Climate and Energy started in 2008 with the aim to gather local governments to achieving the EU climate and energy targets. Municipalities or groups of them (called "signatories") can adhere to CoM voluntarily and sign a declaration stating that they plan to obtain specific targets of CO_2

www.witpress.com, ISSN 1743-3541 (on-line)
doi:10.2495/EQ180041

emissions reduction through specific local actions that are described in the Sustainable Energy Action Plan (SEAP) [22]. CoM initiative is part of the world climate change networks such as the Conference of Parties 21, "C40 cities", the ICLEI – Local Governments for Sustainability [23] and the Smart Cities Initiative [24].

Several studies on the topic of the CoM initiative have been published so far, analysing in detail different scales of the initiative, from national [22], [25], to regional [26]–[31] and local [32]–[35]. The comparison between municipalities of different countries has also been considered [36]–[39].

Local authorities play a leading role in the application of the SEAPs [40]–[42] and their economic investments to the global targets of the Paris agreement [43] are supported by citizens [44], [45] even if issues and barriers are still remaining [46]. Opportunities to link these actions to a circular economy vision come from projects like BIOREGIO [47].

Support to GHG reduction at municipal level has recently been afforded by recent studies [48], [49] through tools [50]–[53], indicators [54] and specific actions on several public sectors [55]–[59] that helps administrations improving the effectiveness of the action plans. Despite the broad interest in the initiative and the increasing dimension of the population involved in this initiative (250+ million inhabitants and 7700+ signatories), the status of implementation of the Covenant of Mayors by country is still lacking and an overview of the whole initiative is missing. The focus of the work is to provide a review of the first 10 years of the initiative and the status of implementation of the CoM in Italy, that is actually the country with the largest population involved by this initiative.

2 THE COVENANT OF MAYORS IN THE DECADE 2008–2017

Since 2008, CoM initiative provided an increasing number of signatories, both local and regional authorities, reaching in the period 2009–2013 the average of 963 new signatories per year (see Fig. 1(a)). Until 2015, majority of the new signatories fixes the targets of the SEAP to the year 2020. Starting from year 2016, the target of CO_2 reduction was updated with the concept of climate mitigation and adaptation with a 2030 horizon. Action plan in this case is called Sustainable Energy and Climate Action Plan (SECAP).

From 2014, signatories started publishing the so-called Progress Plans (see Fig. 1(b)). Two types can be published: (i) the Monitoring Report, related to the mitigation and adaptation actions adopted, every two years; (ii) the Emission Inventory, related to resilience to climate change, reduction of emission and energy consumption, every four years.

Progress plans cover up to seven sectors in which specific targets are identified: industry, local electricity production, local heat cold production, municipal buildings equipment facilities, public lighting, residential buildings, tertiary buildings, transport; an eighth sector, called others, includes the remaining local fonts of energy consumption.

Despite the first signatures in 2008, we see in Fig. 1(b) that the first progress plans have been published in 2014, with a delay of six years. In the decade 2008–2017 a total of 1657, 105 and 64 progress plans have been published considering respectively targets 2020, adaptation and 2030. If we analyse the ratio between the total progress plan and the total action plans, we find that for 2020 target we have a coverage of 52.7%, while for adaptation and 2030 targets, the coverage rises up respectively to 105.0% and 116.4%. Adaptation and 2030 target from this analysis appear to be much better monitored respect to 2020 action plans. Causes of these different coverages need to be deeper analysed since implementation progresses play a fundamental role on the global achievement of GHG reduction [58].

CoM is presently the world's largest initiative that considers actions on energy saving and on local climate. CoM started in Europe in 2008, including in October 2010 already more than 2000 signatories.

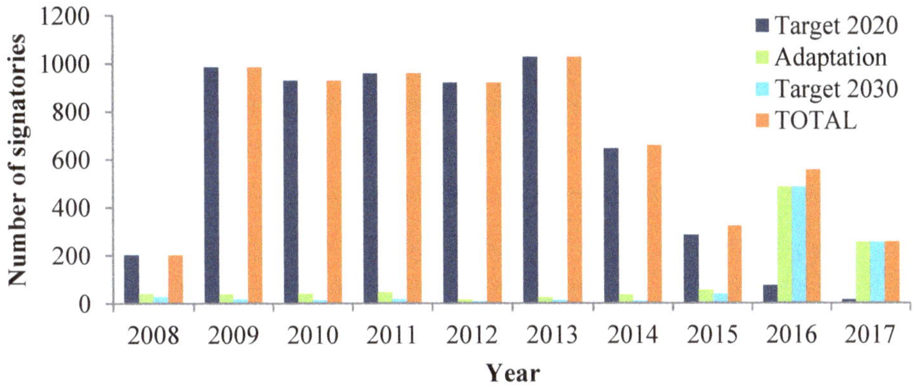

(a)

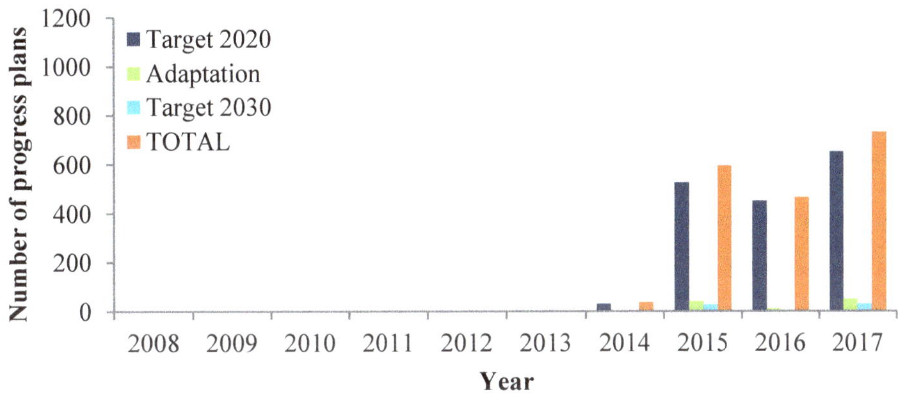

(b)

Figure 1: (a) Signatories of SEAP-SECAP in the period 2008–2017; (b) Progress plans in the period 2008–2017. Colours of the histogram represent the target of the SAEP-SECAP: 2020 (dark blue), adaptation (green), 2030 (light blue). Total number of signatories per year is represented by orange histogram.

The European Commission (EC) decided to broader the initiative in the EU Eastern Parnership countries (2011), to the European Neighbourhood South Region (2012) and Sub-Saharan Africa, America, Japan, India, China and South-East Asia (2015). Mayors Adapt initiative and Compact of Mayors initiative joined together with CoM respectively in 2015 and 2016, resulting in the present Global Covenant of Mayors for Climate and Energy.

In Fig. 2 we present the list of 53 countries with at least one local signatory of SEAP-SECAP in the period 2008–2017 and their relative population involved in the CoM initiative. Fig. 2 shows with different colours the European member states (EU28, in blue) and the non-EU28 states (in red). For the period 2008–2017, EU28 countries sum up 79.7% of the whole population involved in the initiative. The upper 10% percentile, that covers 5 countries of the list, includes Italy, Spain, United Kingdom, Germany and France. All these countries are in the EU28 and the sum of the population of their signatories covers 53.7% of the whole signatories. The non-EU28 countries with the larger population involved in the initiative are Ukraine and Turkey, both with more than 10 million inhabitants.

Italy is presently the nation with the largest population involved in the initiative, representing by itself 17.3% of the total. In the next paragraph we describe the Italian experience and its status of implementation.

3 COVENANT OF MAYORS IN ITALY

In the period 2008–2017 the number of signatories (single authority and group of authorities) of the CoM initiative in Italy is 3172. Considering the single local authority covered by SEAPs in Italy, we have 3518 signatories with 2020 targets (43.99% of the total), 252 signatories with 2030 targets (3.15%) and 314 signatories with adaptation target (3.93%), on a total of 7998 local authorities in the country. Total population covered by CoM initiative is 41.8 million citizens (68.91%). We can distinguish 4 phases: (i) 2008–2010 in which almost no signatories are present, (ii) 2011 represents the rising part of the distribution (iii) 2012–2015 with high number of signatories (75% of the total) and (iv) 2016–2017 in which a decreasing trend is present. 2 peaks of signatures are present in 2013 and 2015, with more than 700 signatories per year.

Distribution of specific targets identified in the SEAPs in Italy in the period 2008–2017 is shown in Fig. 3. We can distinguish three main classes of targets in term of term of their presence in SEAPs: (A) high-presence targets are represented by municipal buildings equipment facilities, local electricity production and public lighting, (B) intermediate-presence targets include transport, residential buildings and others and (C) low-presence targets include local heat-cold production, tertiary buildings equipment facilities and industry.

Class A includes the two main sources of public energy consumption of a local authority: municipal buildings and public lighting. Local electricity production is in class A, while local heat-cold production is in class C (respectively representing 20.84% and 3.46% of the total).

Private buildings (residential, industry and tertiary) sum 13% of the total even if these sources, together with the source due to transport, commonly represent the majority of GHG emission of local government areas.

In terms of the year selected to reach the target of the SEAP signed, almost all the signatories opted for 2020 (3143, 98.71% of the total), while only 28 signatories in the period 2008–2017 planned targets with 2030 target year. 2020 targets are present in the signatures in the period 2010-2017, while 2030 targets appear only starting from 2016.

GHG reduction percentage is freely set by the signatories. Reference year for the reduction target is 1990. As already pointed out (e.g. in [22]), up to now rarely 1990 was chosen as reference year mainly due to difficulty for the local administration to evaluate data referred to the year 1990. More likely, year of the Baseline Emission Inventory (BEI) is between 2008 and 2010, e.g. in the Portuguese case [22], resulting in some cases in a more ambitious process to reach the selected target, even if this cannot be always the case [60], [61]. Majority of Italian local administration opted for GHG reduction target in the range 20%–30% (79.53% of the total), while 12.54% and 4.40% respectively fixed the GHG emission reduction target in the ranges 30%–40% and 40%–50%. 103 signatories (2.98% of the total) fixed reduction targets over 50%, in some cases also assuming to become completely carbon-neutral by 2050.

Majority of signatories and population (respectively 97.44% and 60.32% of the total) involved in the CoM initiative in Italy are part of small or extra small administrations (population respectively below 50000 or 10000 inhabitants). Different distributions in term of population covered by CoM initiative are present.

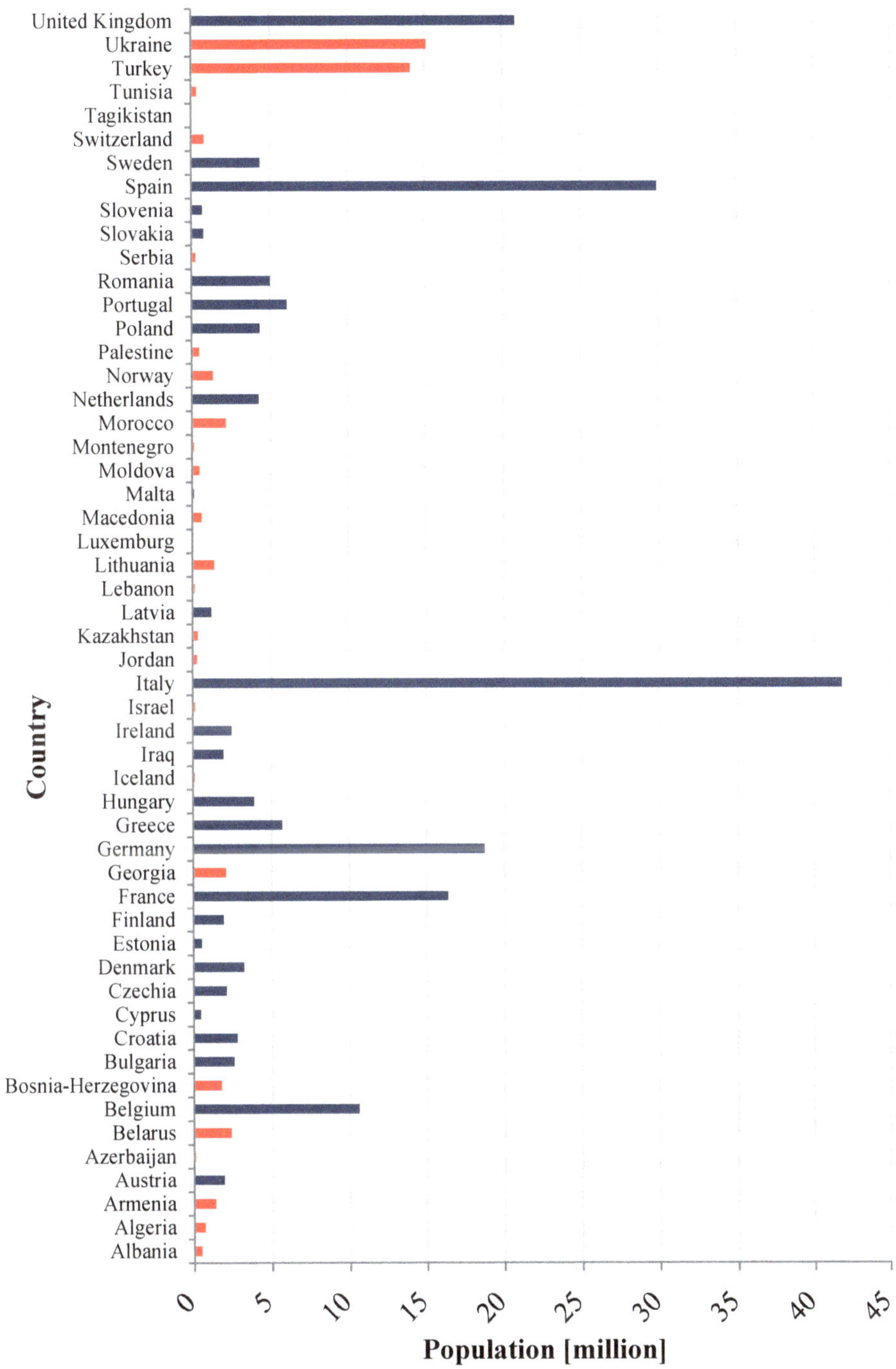

Figure 2: Population per Country covered by Action Plans. Colours of the histogram represent the EU28 countries (blue) and extra EU28 countries (red) respectively.

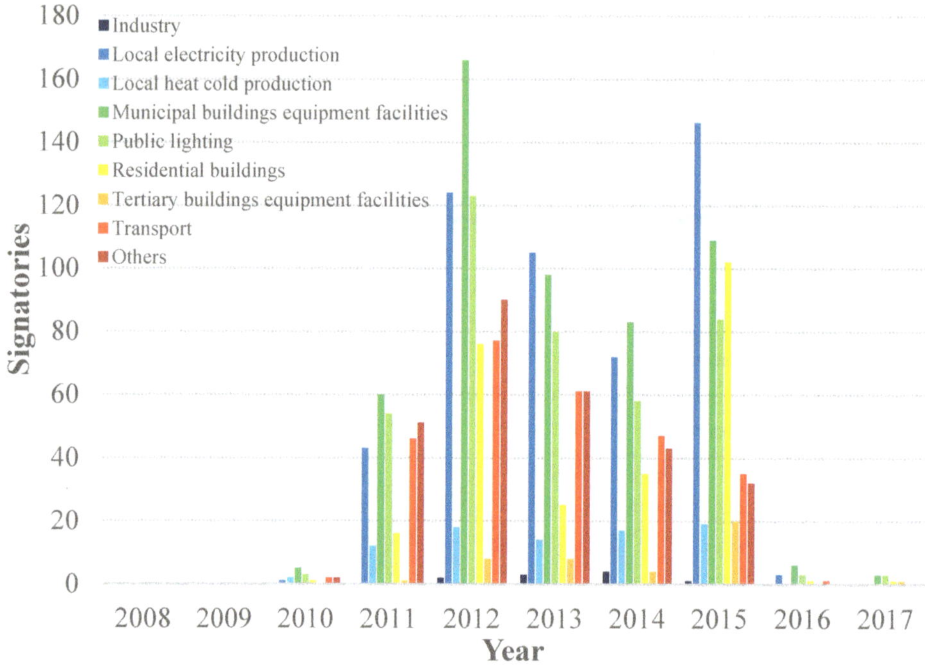

Figure 3: Sectors for which a specific action is assumed in SEAPs in Italy, in the period 2008–2017.

In Fig. 2 we show the territorial coverage assigned to each municipality in Italy participating to CoM distinguished per target: (a) 2020 target, (b) adaptation and (c) 2030 target.

2020 targets cover 141190 km^2 (46.7% of the total Italian territory), while adaptation and 2030 target cover respectively 10119 km^2 and 13774 km^2 (3.4% and 4.6%).

We consider for the macro-analyses 5 Italian administrative units (AU) as defined by EU first level Classification of Territorial Units for Statistics: North West (Aosta Valley, Liguria, Lombardy, Piedmont), North East (Emilia-Romagna, Friuli-Venezia Giulia, Trentino-Alto Adige/Südtirol, Veneto), Centre (Lazio, Marche, Tuscany, Umbria), South (Abruzzo, Apulia, Basilicata, Calabria, Campania, Molise), Islands (Sardinia, Sicily). In three AU (North West, North East and Islands) the coverage of 2020 mitigation target is over 60% (respectively 62.84%, 66.07% and 66.95%), while Centre and Islands population coverage is lower (respectively 30.46% and 47.51%). The two regions with the lowest coverage, Lazio and Umbria, are part of the Centre AU, with a coverage respectively of 23.76% and 23.16%, while the two regions with the highest coverage, Abruzzo and Emilia-Romagna (respectively with 79.56% and 72.49%), are part of the South and North East AU.

4 CONCLUSIONS

In this paper we presented a review of the first 10 years of the Covenant of Mayors initiative in terms of global parameters like the number of signatories and SEAP-SECAP published, countries and population involved, temporal horizon of the targets, status of implementation

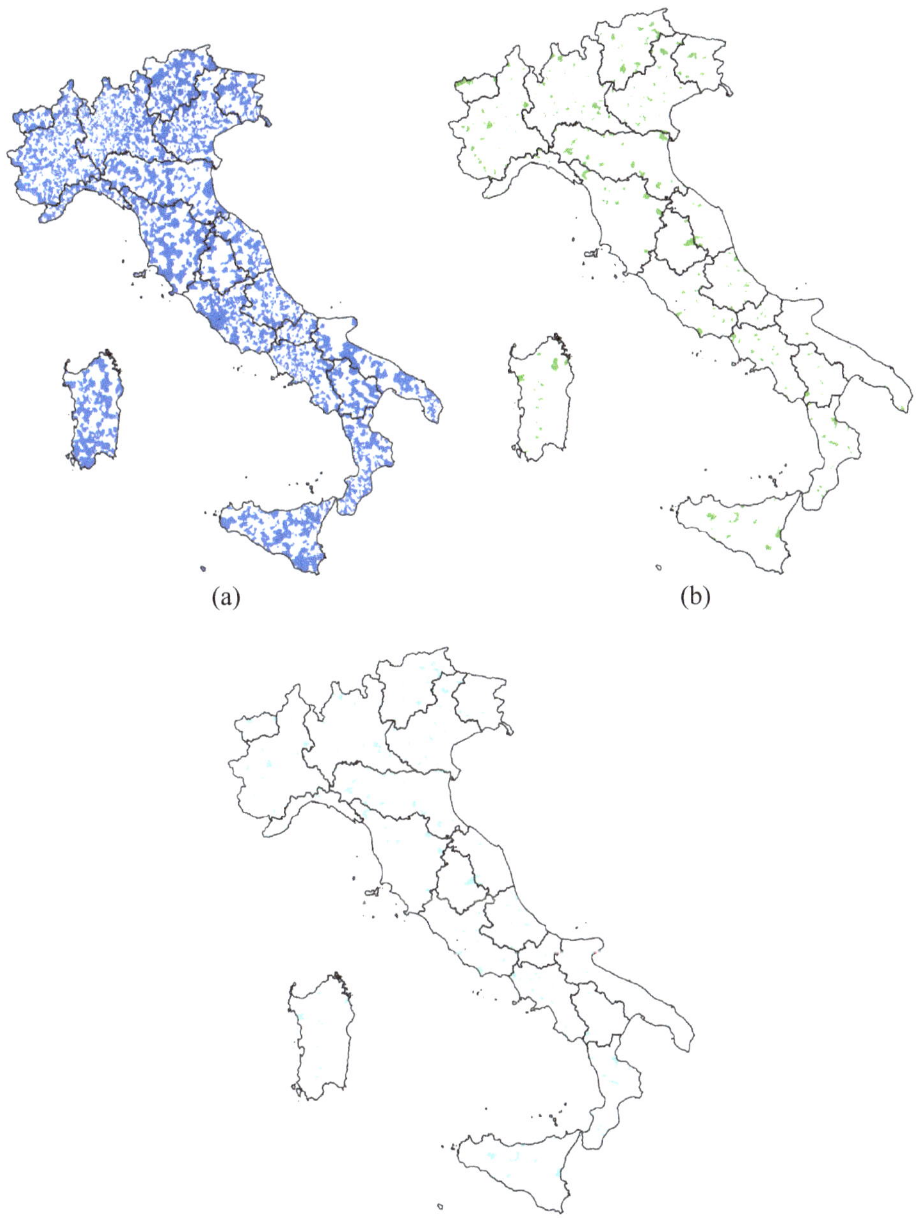

(a) (b) (c)

Figure 4: Status of implementation of the Covenant of Mayors in Italy, showing the administrative areas covered by signatories: Target 2020 (a), in blue; Adaptation (b), in green; Target 2030 (c), in light blue.

of the Progress Plans. Success in terms of signatories and population involved by CoM voluntary initiative in its first decade of activity underlines the increasing awareness at local scale of the global goal of low-carbon economy. Considering EU28, territorial coverage of initiative is not homogeneous per country, with a higher density in the Mediterranean area (south of Spain and Italy in particular) and in Belgium.

Furthermore, we showed the status of implementation of the CoM initiative in Italy, that is presently the country with the larger population involved. In terms of specific targets, high-presence of the categories named (i) municipal buildings equipment facilities, (ii) local electricity production and (iii) public lighting emphasises the role of the signatories, that are public local authorities or group of them. Private contribution to the local targets due to e.g. residential buildings, tertiary buildings and industries is less represented in the SEAP-SECAP, even if they largely contribute, together with transport, to the total of GHG emissions.

Despite EU encourages climate actions with local emphasis, sometimes a limited presence of technical staff supporting local authorities that is appropriately trained on this specific topic could limit the development of targets and consequently the monitoring phase.

Local sustainable energy production, in the more general view of a local circular economy, can positively contribute to reach the goals fixed by EU and in order to improve the resilience of Earth's climate, but it is necessary to constantly monitor the whole of the actions and also to have an estimate of the emissions (e.g. BEI here described) of local authorities that are actually excluded from the initiative in order to give an overall opinion in terms of percentage of GHG emissions.

ACKNOWLEDGEMENT

Thanks are due for the financial support to Municipality of Oppeano (Italy).

REFERENCES

[1] Hurrell, J.W., Decadal trends in the North Atlantic oscillation: Regional temperatures and precipitation. *Science,* **269**(5224), pp. 676–679, 1995. DOI: 10.1126/science.269.5224.676.

[2] Zachos, J., Pagani, H., Sloan, L., Thomas, E. & Billups, K., Trends, rhythms, and aberrations in global climate 65 Ma to present. *Science,* **292**(5517), pp. 686–693, 2001. DOI: 10.1126/science.1059412.

[3] Walther, G. et al., Ecological responses to recent climate change. *Nature,* **416**(6879), pp. 389–395, 2002.

[4] Parmesan, C. & Yohe, G., A globally coherent fingerprint of climate change impacts across natural systems. *Nature,* **421**(6918), pp. 37–42, 2003. DOI: 10.1038/nature01286.

[5] Rockström, J. et al., A safe operating space for humanity. *Nature,* **461**(7263), pp. 472–475, 2009.

[6] Thomas, C.D. et al., Extinction risk from climate change. *Nature,* **427**(6970), pp. 145–148, 2004.

[7] Butchart, S.H.M. et al., Global biodiversity: Indicators of recent declines. *Science,* **328**(5982), pp. 1164–1168, 2010.

[8] Scheffer, M., Carpenter, S., Foley, J.A., Folke, C. & Walker, B., Catastrophic shifts in ecosystems. *Nature,* **413**(6856), pp. 591–596, 2001. DOI: 10.1038/35098000.

[9] Rogelj, J. et al., Paris Agreement climate proposals need a boost to keep warming well below 2°C. *Nature,* **534**(7609), pp. 631–639, 2016. DOI: 10.1038/nature18307.

[10] Schleussner, C.-F. et al., Differential climate impacts for policy-relevant limits to global warming: The case of 1.5°C and 2°C. *Earth System Dynamics,* **7**(2), pp. 327–351, 2016. DOI: 10.5194/esd-7-327-2016.

[11] Schleussner, C.-F. et al., Science and policy characteristics of the Paris Agreement temperature goal. *Nature Climate Change,* **6**(9), pp. 827–835, 2016. DOI: 10.1038/nclimate3096.
[12] The EU Strategy on adaptation to climate change; European Union. ec.europa.eu/clima/sites/clima/files/docs/eu_strategy_en.pdf. Accessed on: 13 Apr. 2018.
[13] Guidelines on developing adaptation strategies, Commission Staff Working Document; European Union, SWD 134 (2013). ec.europa.eu/clima/sites/clima/files/adaptation/what/docs/swd_2013_134_en.pdf. Accessed on: 11 Apr. 2018.
[14] Principles and recommendations for integrating climate change adaptation considerations under the 2014–2020 rural development programmes, Commission Staff Working Document; European Union, SWD 139 (2013). ec.europa.eu/clima/policies/adaptation/what/docs/swd_2013_139_en.pdf. Accessed on: 24 Apr. 2018.
[15] Adapting infrastructure to climate change, Commission Staff Working Document; European Union, SWD 139 (2013). ec.europa.eu/clima/sites/clima/files/adaptation/what/docs/swd_2013_137_en.pdf. Accessed on: 27 Apr. 2018.
[16] Low-carbon economy. https://ec.europa.eu/clima/policies/strategies/2050_en. Accessed on: 30 Mar. 2018.
[17] Pichler, P.-P., Zwickel, T., Chavez, A., Kretschmer, T., Seddon, J. & Weisz, H., Reducing urban greenhouse gas footprints. *Scientific Reports,* **7**(1), p. 14659, 2017. DOI: 10.1038/s41598-017-15303-x.
[18] Kona, A. et al., The Covenant of Mayors in Figures and Performance Indicators: 6-Year Assessment. *JRC Science and Policy Reports. European Commission, Publications Office of the European Union,* 2015.
[19] Lombardi, M., Pazienza, P. & Rana, R., The EU environmental-energy policy for urban areas: The Covenant of Mayors, the ELENA program and the role of ESCos. *Energy Policy,* **93**, pp. 33–40, 2016. DOI: 10.1016/j.enpol.2016.02.040.
[20] Cosmi C., Salvia M., Di Leo S., Pietrapertosa F. & Loperte S., Interregional Cooperation as a Key Tool for the Achievement of Strategic-Energy and Climate Targets: The Experience of the INTERREG IVC RENERGY and SEE RE-SEEties Projects. *Smart and Sustainable Planning for Cities and Regions. SSPCR 2015. Green Energy and Technology*, eds. A. Bisello, D. Vettorato, R. Stephens & P. Elisei, pp. 335–352, 2017.
[21] Gargiulo, M. et al., An integrated planning framework for the development of sustainable and resilient cities—the case of the InSMART project. *Procedia Engineering,* **198**, pp. 444–453, 2017. DOI: 10.1016/j.proeng.2017.07.099.
[22] Coelho, S., Russo, M., Oliveira, R., Monteiro, A., Lopes, M. & Borrego, C., Sustainable energy action plans at city level: A Portuguese experience and perception. *Journal of Cleaner Production,* **176**, pp. 1223–1230, 2018. DOI: 10.1016/j.jclepro.2017.11.247.
[23] ICLEI—Local Governments for Sustainability, Local Solutions to Global Challenges, ICLEI Bonn Center, Bonn. http://archive.iclei.org/fileadmin/user_upload/documents/Global/About_ICLEI/brochures/ICLEI-intro-2009.pdf. 2010.
[24] European Commission, Investing in the Development of Low Carbon Technologies (SET-plan)—Communication from the commission to the European Parliament, the Council, the European Economic and Social Committee and the Committee of the Regions, COM (2014), vol. 15. European Commission: Brussels, 2009.
[25] Batas Bjelic, I. & Ciric, R.M., Optimal distributed generation planning at a local level—A review of Serbian renewable energy development. *Renewable and Sustainable Energy Reviews,* **39**, pp. 79–86, 2014. DOI: 10.1016/j.rser.2014.07.088.

[26] Oliver-Solà, J., Armero, M., De Foix, B.M. & Rieradevall, J., Energy and environmental evaluation of municipal facilities: Case study in the province of Barcelona. *Energy Policy,* **61**, pp. 920–930, 2013. DOI: 10.1016/j.enpol.2013.06.053.
[27] Famoso, F., Lanzafame, R., Monforte, P. & Scandura, P.F., 2015. Analysis of the covenant of mayors initiative in sicily. *Energy Procedia,* **81**, pp. 482–492, 2015. DOI: 10.1016/j.egypro.2015.12.122.
[28] Mahbub, M.S., Viesi, D. & Crema, L., Designing optimized energy scenarios for an Italian Alpine valley: the case of Giudicarie Esteriori. *Energy,* **116**, pp. 236–249, 2016. DOI: 10.1016/j.energy.2016.09.090.
[29] Matak, N., Krajačić, G. & Pilato, A.M., Integrating sustainable energy action plans for island municipalities: Case study of Korcula. *Thermal Science,* **20**(4), pp. 1037–1048, 2016. DOI: 10.2298/tsci151127109m.
[30] Pablo-Romero, M.D.P., Pozo-Barajas, R. & Sánchez-Braza, A., Analyzing the effects of Energy Action Plans on electricity consumption in Covenant of Mayors signatory municipalities in Andalusia. *Energy Policy,* **99**, pp. 12–26, 2016. DOI: 10.1016/j.enpol.2016.09.049.
[31] Roca, A.V., The promotion of photovoltaic for self-consumption in the local public administration: The case of the Barcelona Provincial Council. *Revista de Obras Publicas,* **164**(3584), pp. 86–91, 2017.
[32] Lombardi, M., Rana, R., Pazienza, P. & Tricase, C., The European policy for the sustainability of urban areas and the covenant of mayors' initiative: A case study. *Pathways to Environmental Sustainability: Methodologies and Experiences,* eds. R. Salomone & G. Saija, Springer-Verlag: Berlin and New York, pp. 183–192, 2014.
[33] Schenone, C., Delponte, I. & Pittaluga, I., The preparation of the sustainable energy action plan as a city-level tool for sustainability: The case of Genoa. *Journal of Renewable and Sustainable Energy,* **7**(3), p. 033126, 2015. DOI: 10.1063/1.4921723.
[34] Nuss-Girona, S., Llausàs, A., Figueras, J. & Morera, S., The SEAP in the city of Girona, a crossroads between boldness and pragmatism. *Local Environment,* **21**(4), pp. 476–503, 2016. DOI: 10.1080/13549839.2014.974150.
[35] Scorza, F., Santopietro, L., Giuzio, B., Amato, F., Murgante, B. & Casas, G.L., Conflicts between environmental protection and energy regeneration of the historic heritage in the case of the city of Matera: Tools for assessing and dimensioning of sustainable energy action plans (SEAP). *Lecture Notes in Computer Science,* **10409**, pp. 527–539, 2017. DOI: 10.1007/978-3-319-62407-5_37.
[36] Amorim, E.V., Sustainable energy action plans: project management intercomparison. *Procedia Technologies,* **16**, pp. 1183–1189, 2014. DOI: 10.1016/j.protcy.2014.10.133.
[37] Croci, E., Lucchitta, B., Janssens-Maenhout, G., Martelli, S. & Molteni, T., Urban CO2 mitigation strategies under the Covenant of Mayors: An assessment of 124 European cities. *Journal of Cleaner Production,* **169**, pp. 161–177, 2017. DOI: 10.1016/j.jclepro.2017.05.165.
[38] Haupt W., European municipalities engaging in climate change mitigation and adaptation networks: Examining the case of the covenant of mayors. *Resilience-Oriented Urban Planning. Lecture Notes in Energy,* eds. Y. Yamagata & A. Sharifi, Springer International Publishing: Cham, **65**, pp. 93–110, 2018.
[39] Pietrapertosa, F., Khokhlov, V., Salvia, M. & Cosmi, C., Climate change adaptation policies and plans: A survey in 11 South East European countries. *Renewable and Sustainable Energy Reviews,* **81**, pp. 3041–3050, 2018. DOI: 10.1016/j.rser.2017.06.116.

[40] Lee, T., Local energy agencies and cities' participation in translocal climate governance. *Environmental Policy and Governance,* **28**(3), pp. 131–140, 2018. DOI: 10.1002/eet.1798.
[41] Melica, G., Bertoldi, P., Kona, A., Iancu, A., Rivas, S. & Zancanella, P., Multilevel governance of sustainable energy policies: The role of regions and provinces to support the participation of small local authorities in the Covenant of Mayors. *Sustainable Cities and Society,* **39**, pp. 729–739, 2018. DOI: 10.1016/j.scs.2018.01.013.
[42] Pablo-Romero, M.D.P., Pozo-Barajas, R. & Sánchez-Braza, A., Analyzing the effects of the benchmark local initiatives of Covenant of Mayors signatories. *Journal of Cleaner Production,* **176**, pp. 159–174, 2018. DOI: 10.1016/j.jclepro.2017.12.124.
[43] Bertoldi, P., Kona, A., Rivas, S. & Dallemand, J.F., Towards a global comprehensive and transparent framework for cities and local governments enabling an effective contribution to the Paris climate agreement. *Current Opinion in Environmental Sustainability,* **30**, pp. 67–74, 2018. DOI: 10.1016/j.cosust.2018.03.009.
[44] Pablo-Romero, M.P., Sánchez-Braza, A. & Manuel González-Limón, J., Covenant of Mayors: Reasons for Being an Environmentally and Energy Friendly Municipality. *Review of Policy Research,* **32**(5), pp. 576–599, 2015. DOI: 10.1111/ropr.12135.
[45] Martelli, S. et al., Do voters support local commitments for climate change mitigation in Italy? *Ecological Economics,* **144**, pp. 27–35, 2018. DOI: 10.1016/j.ecolecon.2017.06.035.
[46] Christoforidis, G.C., Chatzisavvas, K.C., Lazarou, S. & Parisses, C., Covenant of Mayors initiative. Public perception issues and barriers in Greece. *Energy Policy,* **60**, pp. 643–655, 2013. DOI: 10.1016/j.enpol.2013.05.079.
[47] www.interregeurope.eu/bioregio/.
[48] Cellura, M., Cusenza, M.A. & Longo, S., Energy-related GHG emissions balances: IPCC versus LCA. *Science of the Total Environment,* **628–629**, pp. 1328–1339, 2018. DOI: 10.1016/j.scitotenv.2018.02.145.
[49] Martire, S., Mirabella, N. & Sala, S., Widening the perspective in greenhouse gas emissions accounting: The way forward for supporting climate and energy policies at municipal level. *Journal of Cleaner Production,* **176**, pp. 842–851, 2018. DOI: 10.1016/j.jclcpro.2017.12.055.
[50] Kyriakarakos, G., Patlitzianas, K., Damasiotis, M. & Papastefanakis, D., A fuzzy cognitive maps decision support system for renewables local planning. *Renewable and Sustainable Energy Reviews,* **39**, pp. 209–222, 2014. DOI: 10.1016/j.rser.2014.07.009.
[51] Cochran, I., Variant 1: Region/city geographical inventories. V. Bellassen, & N. Stephan, *Accounting for carbon: Monitoring, reporting and verifying emissions in the climate economy. Accounting for Carbon: Monitoring, Reporting and Verifying Emissions in the Climate Economy*, Cambridge University Press: Cambridge, pp. 72–103, 2015.
[52] Marinakis, V., Papadopoulou, A.G., Doukas, H. & Psarras, J., A web tool for sustainable energy communities. *International Journal of Information and Decision Sciences,* **7**(1), pp. 18–31, 2015. DOI: 10.1504/ijids.2015.068115.
[53] Marinakis, V., Xidonas, P. & Doukas, H., A modelling framework for the forecasting of energy consumption and CO2 emissions at local/regional level. *International Journal of Global Energy Issues,* **39**(6), pp. 444–460, 2016. DOI: 10.1504/ijgei.2016.079374.
[54] Zanni, D., Righi, A., Dalla Mora, T., Peron, F. & Romagnoni, P., The Energy improvement of school buildings: Analysis and proposals for action. *Energy Procedia,* **82**, pp. 526–532, 2015. DOI: 10.1016/j.egypro.2015.11.865.

[55] Dall'O', G., Galante, A. & Pasetti, G., A methodology for evaluating the potential energy savings of retrofitting residential building stocks. *Sustainable Cities and Society,* **4**(1), pp. 12–21, 2012. DOI: 10.1016/j.scs.2012.01.004.
[56] Dall'O', G., Galante, A., Sanna, N. & Miller, K., On the integration of leadership in energy and environmental design (LEED)® ND protocol with the energy planning and management tools in Italy: Strengths and weaknesses. *Energies,* **6**(11), pp. 5990–6015, 2013. DOI: 10.3390/en6115990.
[57] Berghi, S., Energy use in Urban Transport sector within the Sustainable Energy Action Plans (SEAPs) of three Italian Big Cities. *Energy Procedia,* **126**, pp. 414–420, 2017. DOI: 10.1016/j.egypro.2017.08.193.
[58] Cipriano, X., Gamboa, G., Danov, S., Mor, G. & Cipriano, J., Developing indicators to improve energy action plans in municipalities: An accounting framework based on the fund-flow model. *Sustainable Cities and Society,* **32**, pp. 263–276, 2017. DOI: 10.1016/j.scs.2017.03.004.
[59] Hsu, A., Cheng, Y., Weinfurter, A., Xu, K. & Yick, C., Track climate pledges of cities and companies. *Nature*, **532**(7599), pp. 303–306, 2016. DOI: 10.1038/532303a.
[60] Kona, A. et al., *Covenant of Mayors: Greenhouse Gas Emissions Achievement and Projections*, Publications Office of the European Union: Luxembourg, EUR 28155 EN, 2016.
[61] Olivier, J., Janssens.Maenhout, G., Mountean, M. & Peters, J., Trends in global CO_2 emissions: 2015 Report. Background Studies. PBL Netherlands Environmental Assessment Agency; Ispra: European Commission, Joint Reasearch Centre, The Hague, 2015.

ENERGY AND THE CITY

ALIYU B. DAHIRU, SIRAJDIN ADESHINA & MOHAMMED K. HALIDU
Department of Computer Engineering, Federal Polytechnic Mubi, Adamawa State, Nigeria

ABSTRACT

By concept and analysis, cities are known to be the center stage for the consumption of energy. Their formations are directly related to the building arrangement and road networks, which are factors that are directly determined by the ministry of urban planning and development. This research work is designed to establish the relative magnitudes of energy consumption in buildings in comparison to transportation network, with a clear knowledge that points out the relationships between buildings and transportation in the context of climate change and social inclination in cities. The main objective of the paper is to examine the energy trends and implications of urbanization by making reference to building density and proffer strategic solutions.
Keywords: cities, buildings, transportation, urban planning, energy consumption, climate change, social inclination, density, city texture.

1 INTRODUCTION

Energy as an entity is one aspect of life that performs the most important role in the sustainability of economic growth, progress, development, poverty eradication, security and comfort of any nation. In an effort to meet the climatic and environmental policy goals in Nigeria, several measures are being analyzed. The processes of analyzing these objectives lead to key challenges relating to the planning of energy related urban infrastructures because planning encompasses the determination of locations for new power generating facilities [1]. Steady power supply is an important issue for all nations today. Similarly, the economic growth and future of any nation specifically depends on the availability of sources of energy that are available, reasonable, and environmentally friendly. Security, climate change, and public health are closely interrelated with energy [2].

The standard of living of a given country can be directly related to the per capita energy consumption and the per capita energy consumption is a measure of the per capita income as well as a measure of the wealth of a nation [3]. Research has also shown that the recent global energy crisis is as a result of the massive population growth and the increase in the living standard of societies as a whole [4]. The desire for people to strongly incline themselves to living in the cities gives rise to both positive and negative effects on energy consumption globally [5]. On one hand it causes the increase of cultural level, the creation of new job opportunities and improvement of economic conditions. On the other hand, concentration in cities increased traffic jam, carbon dioxide, greenhouse gases emissions and waste disposal which threaten health care conditions.

The size of a city drives energy consumption hence the need for territory redevelopment and obtainability of adequate infrastructures. In this development, the ideas of save the earth and people's health gives rise to the emergence of smart cities since smart cities are able to solve urban issues paying critical attention to climatic change. This is the reason why the idea of smart growth has begun to spread in the nineties. It involves a community-based reaction aimed at solving transportation issues, school overcrowding, carbon dioxide and greenhouse gas emission, lack of open space and the cost of skyrocketing public facilities [6]. The sustainability of urban development concepts give rise to the awareness, production and utilization of resources needed for industrial, residential, commercial, transportation, and

WIT Transactions on Ecology and the Environment, Vol 222,
www.witpress.com, ISSN 1743-3541 (on-line)
doi:10.2495/EQ180051

recreational processes [7]. Ecological urban development supports environmental awareness in the application of natural resources in the development of smart cities [8].

As a result, demand for energy in buildings and the transportation sector will rise as the growth of urbanization is expected to continue since urban areas are growing due to economic development and industrialization leading to more people attraction to live in the cities than ever before [9]. This rise in energy demand in both buildings and transport give rise to countless contribution of urban areas to greenhouse gas emissions that led to negative consequences such as climate change and air pollution outlined earlier [10]. To implement mitigation strategies, identifying energy consumption patterns in cities and other urban areas is of paramount importance. Buildings and transport are considered the key contributors to the urban energy demand therefore assessing their consumption is of immense importance [11].

The effective energy consumption of both buildings and transport is closely related as a result of urban layout influencing the building users, mobility and respective annual travel distances which affects the carbon footprint in the transportation sector [12]. Transport networks have an impact on the operational energy consumption of both buildings and transport by carrying individuals and goods between places. Studies of the energy usage of both buildings and transport are important to avoid unplanned result of unviable strategies by the ministry of urban planning and development [13]. Climatic change and energy problems in the world have necessitated constituted authorities and institutions at all levels to optimize energy systems in urban densely populated areas thereby improving the standard of living for its citizen [14].

Energy demands predictions for urban buildings are highly important to energy providers intending to boost the energy supplies and introduce renewable energy sources to the electricity mix considering the potential of buildings so as to contribute towards the reductions in carbon footprints globally [15]. Cities therefore consume approximately 70% of global energy and are projected to house approximately 60% of the global population by 2030 [16]. Nigeria's transportation system consists of road, rail, water, and air subsectors. Ninety percent of the total quantity of premium motor spirit consumed and 40% of the automotive gas oil consumed is attributed to road transportation. 24% of the dual-purpose kerosene consumed goes to aviation sector which is the sole fuel. 10% of the fuel oil consumed and 6% of the automotive gas oil consumed is attributed to water transportation [17].

Road transport is the most widely and commonly used means of transportation system in Nigeria today. This specifically depends on the mode of human settlements which account for more than 90% of the subsector's contribution to the gross domestic product. Road transport activities involve the conveyance of passengers in mass or small numbers, conveyance of livestock, farm products, merchandise and rendering of mobile services [18].

Energy consumption in buildings set a center stage in global discursion towards sustainable development and its broader interconnections with the entire environment. Research has shown that energy consumption of the built environment surpasses that of other sectors including transportation in many parts of the world [19]. In the United State, buildings consume as much as 48% and 60% of total energy and electricity respectively [20]. It has therefore become important to access energy consumption in the built environment in a developing world and follow sensible energy procedures in the short term. It is also vital to advocate for a clear policy as well as institutional contexts on the long term. This study examined energy consumption of office buildings and transportation in Abuja which is Nigeria's federal capital territory and its largest urban center. It also considered the fastest growing city in Africa. Abuja has been branded by the presence of significant government and corporate bodies as well as rampant construction activities in an attempt to elicit the status, performance and the impacts of office buildings on energy in the city.

2 CURRENT ENERGY SITUATION AND DEMAND IN NIGERIA

Africa has been blessed with abundant energy sources such as coal, natural gas, petroleum, solar, hydro, geothermal, and nuclear which can be used to increase its current power production, yet Africa's power sector is predominantly weak. Since the inception of industrial revolution, energy has been the driving force for the development of a modern society. Sub-Saharan Africa's situation is the worst in the world from an electricity point of view and is true because this region has 13% of the world's population but 48% of the share of the global population lives without access to electricity.

More than 640 million people in Sub-Saharan Africa are presently living without access to electricity which represents 70% of the overall population. Moving on the present path will take until 2080 for there to be universal electricity access on the African continent. On the other hand Nigeria is estimated to require 12,800 MW of electricity for the year 2016 but the available capacity is only around 4,500 MW [21]. The importance of access to modern energy is now well understood by governments and significant efforts are being made to address energy access challenges.

The present state of Nigerian electricity reflects a situation of crisis between supply and demand where by industrial growth and socio-economic developmental processes operate below the potential of the economy [22]. It is believed that Nigeria is a custodian of abundant energy resources which is dominated by oil and gas. This availability makes it the mainstay of the country's economy accounting for over 85% of export earnings and government revenues. Nigeria is also the 9th largest exporter of crude oil in the world [23].

Table 1: Nigeria's energy reserve (captured December 2017).

Resource	Reserve	Reserves (Billion tonnes of oil equivalent)	Reserves ($x10^7$) TJ
Crude oil	36.2 Billion barrels	4.896	20.499
Natural gas	166 trillion standard cubic feet	4.465	18.694
Coal and lignite	2.7 billion tonnes	1.882	17.652
Tar sands	31 billion barrels of oil equivalent	4.216	17.652
Subtotal fossil		15.459	64.724
Hydropower, large scale	11,000 MW		0.0341/year
Hydropower, small scale	3,250 MW		0.0101/year
Fuel wood	13,071,464 ha		
Animal waste	61 million tonnes/year		
Crop residue	83 million tonnes/year		
Solar radiation	3.5 to 7.0 KWh/day		
Wind	2 to 4 m/s (annual average) at 10 m in height		

Source: Sunday Olayinka Oyedepo (2012). Energy and sustainable development in Nigeria: the way forward. Energy sustainability and society, a spring open journal.

With the availability or abundance of the afore mentioned energy resources, Nigeria is not in any way supposed to depend on the importation of energy to achieve a sustainable generating capacity adequate enough to target ever increasing economic growth and also have excess power generation to sell to neighboring countries [24]. Despite this abundance of energy resources, less than half of the population enjoy access to electricity and reside in the urban metropolis. Fuel wood precisely remains an important energy resource used mainly for cooking in rural and even parts of the urban areas there by aiding desertification [25]. The dominance of fuel wood over the prevalent sources of energy is eminent. This case scenario indicates the extent of competition in areas totally depending on electricity.

3 ENERGY CONSUMPTION PATTERN IN NIGERIA

African countries and in particular Nigeria are observed to have the lowest degree of energy consumption in the world. Although it is typical of any developing economy to suffer from inadequate supply of energy, Nigeria's case is alarming because of increasing demand on daily basis. Nigeria is rich in conventional energy resources which include oil, natural gas, lignite, and coal. It is also well endowed with renewable energy sources such as wood, solar, hydropower, and wind [26]. The pattern of energy usage in Nigeria's economy is divided into industrial, transport, commercial, agricultural, and household sectors. The household sector accounts for the largest energy usage in the country aiming at 65%. The low level of development in all the other sectors account for this percentage [27]. Table 2 below shows that Nigeria has been ranked the number 68 in terms energy consumption by the CIA world fact book 2017.

4 ENERGY CONSUMPTION IN BUILDING IN ABUJA

Abuja is the new capital city of Nigeria. The idea was conceived in the 1970s but the realization and subsequent relocation from Lagos took effect in 1991 by the then military ruler General Babangida. It is located within Latitude of 7° 25' N and 9° 20' North and Longitude of 5° 45' E and 7° 39' E of Greenwich Meridian placing the city exactly at the center of the country and almost at the same distance from anywhere to all its boundaries as shown in Fig. 1. Being at the center it has Kaduna state to the north, Nasarawa state to the east, Kogi state to the south-west and Niger state to the west. The map of Abuja indicating the first and second phases of its development is shown in Fig. 2.

Table 2: Consumption of electricity by the countries of the world.

Rank	Country	Total (Million KWh/year	Per capita (KWh/year)	Date
1	China	5,523,000	4,074	2014 est
2	USA	3,913,000	12,271	2014 est
3	Russia	1,065,000	7,475	2014 est
7	Canada	528,000	15,157	2014 est
8	Brazil	518,000	2,556	2014 est
68	Nigeria	24,000	135	2014 est
69	Hungary	21,550	2189	2014 est
70	Ecuador	21,000	1,341	2014 est

Source: CIA World fact book (2017/01/17). Available at world by map.org/electricity consumption.html.

Table 3: Nigerian energy.

Electricity access	Population without electricity: 95,5000,00
	Electrification – Total population: 45%
	Electrification – Urban Areas: 55%
	Electrification – Rural areas: 37%
Electricity production	29.83 Billion KWh (2015 est)
Electricity consumption	24.57 billion KWh (2015 est)
	Country comparison to the world: 68
Electricity exports	0KWh (2016 est)
	Country comparison to the world: 178
Electricity imports	0KWh (2016 est)
	Country comparison to the world: 181
Electricity – installed generating capacity	10.48 million KW (2015 est)
	Country comparison to the world: 59
Electricity from fossil fuel	80.3% of total installed capacity (2015 est)
	Country comparison to the world: 88
Electricity from nuclear fuels	0% of total installed capacity (2015 est)
	Country comparison to the world: 154
Electricity from hydroelectric plants	19.5% of total installed capacity (2015 est)
	Country comparison to the world: 90
Electricity from other renewable sources	0.2% of total installed capacity (2015 est)
	Country comparison to the world: 163
Refined petroleum products production	70,140 bbl/day (2015 est)
	Country comparison to the world: 72
Refined petroleum products consumption	316,000 bbl/day (2015 est)
	Country comparison to the world: 43
Natural gas production	45.15 billion cu m (2015 est)
	Country comparison to the world: 18
Natural gas consumption	26.86 billion cu m (2015)
	Country consumption to the world: 38
Carbon dioxide from consumption of energy	97 million Mt (2013 est)
	Country comparison to the world: 40

Source: CIA World fact book (2017/01/17). Available at world by map.org/electricity consumption.html.

Table 4: Nigerian transportation.

National air transport system	Number of registered air carriers: 16
	Inventory of registered aircraft operated by air carriers
	Annual passenger traffic on registered air carriers: 3,223,459
	Annual freight traffic on registered air carrier: 22,400,657 mt-kt (2015)
	Airports: 54 (2013)
	Country comparison to the world: 88
	Number of registered air carriers: 16
Railways	Total: 3798 km
	Standard gauge: 293 km 1.435-m gauge (2014)
	Country Comparison to the world: 51
Road ways	Total: 193,200 km
	Paved: 28,980 km
	Unpaved: 164,220 km (2004)
	Country comparison to the world: 29
Waterways	8,600 km (Niger and Benue Rivers and smaller rivers and creeks) (2011)
	Country comparison to the world: 15
Merchant marine	Total: 583
	General cargo: 14
	Oil tanker: 83
	Other; 486 (2017)
Ports and terminals	Major seaport(s): Bonny inshore terminal, Calabar, Lagos
	LNG terminals (export): Bonny Island

Source: CIA World fact book (2017/01/17). Available at world By map.org/electricity consumption.html.

Figure 1: Geographical map of Nigeria.

Figure 2: Map of Abuja.

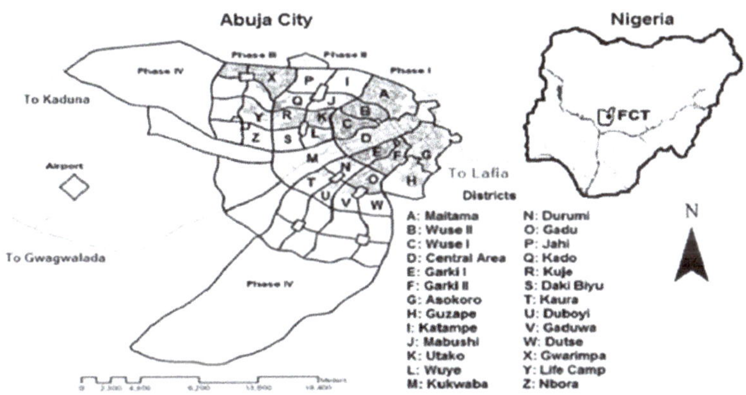

Figure 3: Map showing planned development of Abuja city.

The idea for the creation of the new capital city was prompted by two reasons. The first was the congestion and overcrowding environment in Lagos which was found unbefitting to the country's capital. The second idea was the need to site a capital city in a location that will be neutral to all parties amidst uncertainties propagated by lack of unity and security of the nation from frequent coup plots [28]. The Abuja project design was led by a group of town planners, engineers, foreign architects and also incorporates local architects. At the inception of the project, Abuja was tagged as the biggest project of the time [29]. The development of the main city was divided into four phases as shown in Fig. 3.

The city's land mass is not more than 1% of the total area covered by the country's total land mass with an area of 8,000km^2 and it is increasingly becoming an important urban settlement on the Nigerian landscape [30]. At the completion of Phase 1, the population of Abuja is estimated to be 250,000 from the inception and not more than 3.2 million at the completion of the remaining phases [31]. Contrary to the earlier estimate the population has risen from 400,000 to 1.4 million at the realization of the first phase from 1991 to 2006 showing more than 300% increase [32]. The figures given above are assessment from federal

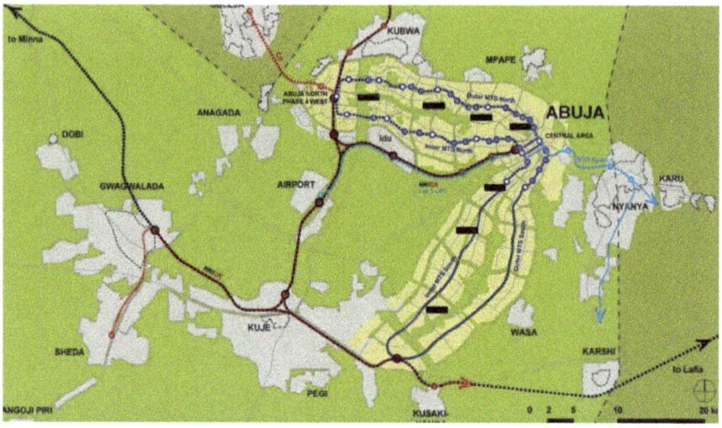

Figure 4: Public transport concept for Abuja.

population commission of Nigeria. The actual figure relates that the city's urban growth rate is said to be between 20–30% annually [33]. The sudden growth rates in Abuja municipal shows a significant effect in the developments on the infrastructure. Regarding energy supply, it was predicted that there was a supply shortage if the infrastructure base was not improved. This give rise to load shading of energy supply to neighboring location so as to give power to the capital [32]. The above sequence highlighted the fact that without significant improvement in the supply of energy dwellers of the city will experience power outages on the completion of the remaining phases. Diesel power generators which generate ample amount of carbon foot print are the only means of alternative source of power supply at the moment [34]. Daily consumption of fuel by generators is estimated to be 12–13 million liters on generators only for the purpose of supplementing power outages that affects all the sectors of the city [35].

5 CITY TEXTURE

Even though insights were gained from assessing urban density in terms of buildings and transport considerations, it is important to divert to the study of the practical or formal intricacy of a city. The study of urban design is a continuous concern at the Martin Centre for Architectural and Urban Studies for over 3 decades [36]. It has already been pointed out that the energy consumption of buildings and transportation is linked to the quality of the urban environment [37].

CONCLUSION

This study assesses the state of buildings and transport in Abuja whereas recognizing the availability of abundant energy resources in country. The study also looked at the energy demand particularly in the urban densely populated environment where electricity is the main source of power supply. Inhabitants are forced to use generators frequently to cover up the wide gap created by power outages. Analyses have shown that commercial and residential buildings account for 37% and 61% consumption of the energy supply respectively. This increase is less than the energy increase in changing from a naturally ventilated office to an air-conditioned office. However, this change is possible if the pollution and noise to the urban environment has been minimized to an appreciable and acceptable low level.

This gives rise to a proportion where the relationship between transport and building energy becomes apparent as cars are the major cause of urban environmental pollution which results in large carbon footprint. It is evident that the overall urban transportation system consumes less energy than the energy consumption of urban infrastructure. It is invariably necessary to note that the effects of reducing the use of private cars in favor of clean and efficient public transport can be significant for building energy use. This confirmed that commercial buildings have a significant impact on the energy supply of a city.

REFERENCES

[1] Kaden, R. & Kolbe, T.H., City-wide total energy demand estimation of buildings using semantic 3d city models and statistical data. *ISPRS Annals of the Photogrammetric, Remote Sensing and Spatial Information Sciences*, vol. II-2/W1, ISPRS 8th 3DGeoInfo Conference & WG II/2 Workshop, Turkey, 2013.

[2] Ramchandra, P. & Boucar, D., *Green Energy and Technology*, Springer: London and New York, 2011.

[3] Rai, G.D., *Non—Conventional Energy Sources*, Khanna Publishers: Deli, 2004.

[4] Sunday, O., *Energy and Sustainable Development in Nigeria: The Way Forward,* Springer open journal, 2012.

[5] Caragliu, A., Del Bo, C. & Nijkamp, P., Smart Cities in Europe, *Journal of Urban Technology,* **18**(2), pp. 65–82, 2011.

[6] Pardo, T. & Taewoo, N., Conceptualizing smart city with dimensions of technology, people, and institutions. *Proceedings of the Twelveth Annual International Conference on Digital Government Research,* New York, pp. 282–291, 2011.

[7] Yigitcanlar, T., Smart cities: An effective urban development and management model? *Australian Planner*, **52**(1), pp. 27–34, 2015. DOI: 10.1080/07293682.2015.1019752.

[8] Pietrosemoli, L. & Monroy, C.R., The impact of sustainable construction and knowledge management on sustainability goals. A review of the Venezuelan renewable energy sector. *Renewable and Sustainable Energy Reviews*, **27**, pp. 683–691. DOI: 10.1016/j.rser.2013.07.056.

[9] Goonetilleke, A., Yigitcanlar, T., Ayoko, G. & Egodawatta, P., *Sustainable urban water environment: Climate, pollution and adaptation,* Cheltenham: Edward Elgar, 2014.

[10] Reinhart, C.F. & Davila, C.C., Urban building energy modeling—A review of a nascent field. *Building and Environment,* **97**, pp. 196–202, 2016. DOI: 10.1016/j.buildenv.2015.12.001.

[11] Hickman, R. & Banister, D., *Transport, Climate Change and the City,* vol. 7, Routledge: London, 2014.

[12] Stephan, A., Crawford, R.H. & De Myttenaere, K., Towards a comprehensive life cycle energy analysis framework for residential buildings. *Energy and Buildings,* **55**, pp. 592–600, 2012. DOI: 10.1016/j.enbuild.2012.09.008.

[13] Barthelemy, M., Bordin, P., Berestycki, H. & Gribaudi, M., Self-organization versus top-down planning in the evolution of a city Nature. *Scientific Reports*, **3**(1), p. 2153, 2013. DOI: 10.1038/srep02153.

[14] Vanolo, A., Smart mentality: The smart city as disciplinary strategy. *Urban Studies,* **51**(5), pp. 883–898, 2013. DOI: 10.1177/0042098013494427.

[15] Pérez-Lombard, L., Ortiz, J. & Pout, C., A review on building energy consumption information. *Energy and Buildings,* **40**(3), pp. 394–398, 2008. DOI: 10.1016/j.enbuild.2007.03.007.

[16] Seto, K.C. & Dhakal, S., Human settlements, infrastructure, and spatial planning in climate change 2014: Mitigation of climate change. *Contribution of Working Group III to the Fifth Assessment Report of the Intergovernmental Panel on Climate Change.* eds O. Edenhofer, R. Pichs-Madruga, Y. Sokona, E. Farahani, S. Kadner, Cambridge University Press, Cambridge, UK, and New York, 2014.
[17] Oniwon, A., *Oil and Gas in Nigeria's National Development: An Assessment.* A presentation at the National Defence College: Abuja, 2011.
[18] Obih, H., *Fuel Distribution and Logistics.* Sub-regional conference on the phase-out of leaded gasoline in Nigeria and neighboring countries: Abuja, Nigeria, 2001.
[19] Brown, G., Architecture Industry; A hidden Culprit in Global Warming. Online. http://architecturalevangelist.com/green-architecture/the-architecture-industry-a-hidden-culprit-to-global-warming.html. 2010.
[20] Architecture2030, Total Energy Consumption in US (By Sector). Online. http://architecture2030.org/the_problem/buildings_problem_why. Accessed on: 18 Dec. 2010.
[21] Castellano, A., Kendall, A., Nikomarov, M. & TarrynSwemmer, T., *Brighter Africa: The growth potential of the sub-Saharan electricity sector.* McKensey & Company, 2015. Online. www.mckinsey.com/industries/electric-powerand-natural-gas/our insights/powering-africa, 2015.
[22] Adeyemi, A.O. & Ayomide, A., Electricity consumption and economic growth in Nigeria. *Journal of Business Management and Applied Economics,* **2**(4), pp. 1–14, 2013.
[23] OPEC, OPEC Bulletin, in Organization of the Petroleum Exporting Countries, 2011.
[24] Koledoye, O., Jumah, A. & Philips, A., The current and future challenges of electricity market in Nigeria in the face of deregulation process. Online. http://theses.covenantuniversity.edu.ng/bitstream/handle/123456789/166/The%20Current%20and%20Future%20Challenges%20of%20Electricity%20Market%20in%20Nigeria%20in%20the%20Face%20of%20Deregulatio n%20Process.pdf?sequence=1. Accessed on: 28 Feb. 2015.
[25] Bugaje, I., Renewable energy for sustainable development in Africa: A review. *Renewable and Sustainable Energy Reviews,* **10**(6), pp. 603–612, 2006. DOI: 10.1016/j.rser.2004.11.002.
[26] Okafor, E. & Joe, U., Challenges to development of renewable energy for electric power sector in Nigeria. *International Journal of Academic Research*, **2**(2), pp. 211–216, 2010.
[27] Energy Commission of Nigeria (ECN) National Energy Policy, Federal Republic of Nigeria: Abuja, 2003.
[28] Salau, A.T., A new capital for Nigeria: Planning, problems and prospects. *Africa Today*, **24**(4): pp. 11–22, 1977.
[29] Take, O., *Abuja, the New Capital of Nigeria, and the Urban Design of Its Central Area,* in *Continuity and Change: Design Strategies for Large-Scale Urban Development*, ed. M.B. Sevcenko, The Aga Khan Program for Islamic Architecture: Cambridge, Massachusetts, 1984.
[30] Mabogunje, A.L., *The Promise, Performance and Prospect in the Review of Abuja Master Plan*, Fountain Publications Ibadan: Abuja, 1999.
[31] Olomola, A.O., *Position Paper of the Nigerian Institute of Town Planners: On the Implementation of the Abuja Master Plan. In the Review of the Abuja Master Plan*, Fountain Publications Ibadan: Abuja, 1999.

[32] Galadima, E.I., *Power Supply and Improvement in the FCT. In the Review of Abuja Master Plan*, Fountain Publications Ltd, Ibadan: Abuja, 1999.
[33] World Bank, Nigeria Country Profile, World Development Indicators and Global Development Finance, World Bank 2011.
[34] Nwaobi, G.C. & Abuja, G., Solving the poverty crisis in Nigeria: An applied general equilibrium approach. *Computational Economics*, 2003.
[35] ECN (Energy Commission of Nigeria), Manufacturers spend N1.8bn weekly on diesel -MAN in Energy Commission of Nigeria. Online. www.energy.gov.ng/index.php?option=com_content&task=view&id=56&Itemid=58. Accessed on: 9 Sept. 2010.
[36] Martin, L. & March, L., *Urban Space and Structures*, Cambridge University Press: Cambridge, 1972.
[37] Steemers, K. & Ratti, C., Informing bioclimatic urban design. *Architecture and Engineering: The Teaching of Architecture for Multidisciplinary Practice,* ed. M. Voyatzaki, European Association for Architectural Education, 1999.

A METHODOLOGY TO SUPPORT DECISIONS TOWARDS ECONOMIC AND ENVIRONMENTAL SUSTAINABILITY IN PUBLIC CONTEXTS: APPLICATION TO HAND-DRYING OPTIONS

MARCO SCHIAVON[1], MARCO RAGAZZI[1], GRAZIANO COLLER[2], NAVARRO FERRONATO[3], VINCENZO TORRETTA[3] & ELENA CRISTINA RADA[3]
[1]Department of Civil, Environmental and Mechanical Engineering, University of Trento, Italy
[2]Department of Economics and Management, University of Trento, Italy
[3]Department of Theoretical and Applied Sciences, University of Insubria, Italy

ABSTRACT
Different hand-drying methods entail different costs and impacts on humans and the environment. This paper presents a methodology to facilitate the decision on the convenience of installing electrical hand driers in place of the conventional toilet paper towels in the restrooms of public places. Specifically, a procedure including both economic and environmental aspects is proposed and the Monte Carlo method is employed to account for the several uncertainties of all the variables involved. From the economic point of view, the number of daily usages of restrooms results as the key variable determining if one option is preferable to the other. From the environmental point of view, the carbon footprint was calculated for four scenarios considering two options for the electric energy grid mix and two options for waste treatment. The comparison between the four scenarios revealed that hand driers may be preferable to paper towels when the waste treatment alternative is landfilling. The results are more uncertain when the waste treatment option is incineration. The integration of economic and environmental aspects reveals as a useful strategy to fully assess the convenience of choosing one option rather than another, without limiting the decision to only one aspect. Additional information on processes and logistics are anyway necessary to reduce the uncertainties of the results.
Keywords: paper towels, hand driers, Monte Carlo method, carbon footprint, energy consumption, waste management, greenhouse gases, road transport, incineration, landfilling.

1 INTRODUCTION

Buildings are important final consumers of energy and, consequently, high direct and indirect contributors to the emissions of greenhouse gases (GHGs) into the atmosphere [1]. Depending on their contexts, the energy demand by buildings is about 25–45% the worldwide energy demand [2], [3]. In public buildings, there is also a general growing concern towards public hygiene, which should be pursued with sustainable strategies to limit both the costs for the society and the impacts of different alternatives on the environment. Besides hand washing, hand drying is an important step in public hygiene, because of the demonstrated higher potentials for carrying microbes (and, thus, infections) on wet hands, in comparison with dry hands [4].

Different hand-drying methods exist: paper towels, cotton rolls and electric hand-driers. The last are warm- or hot-air hand driers (which generate a low-speed warm/hot airflow), high-speed hand driers (which dry users' hands by a jet of unheated air) and the more recent airblade-type hand driers (whose effect is given by a thin layer of unheated air exiting at a higher speed than jet-air hand driers). Different hand-drying methods entail different impacts on humans and the environment. Compared to paper roll/towels, hand driers generally show higher impacts in terms of dispersion of airborne bacteria. In a dated study, Matthews and Newsome [5] found no significant differences between the concentrations of bacteria aerosols in proximity of a hot-hair hand drier and a paper towel dispenser. In a more recent

www.witpress.com, ISSN 1743-3541 (on-line)
doi:10.2495/EQ180061

study, paper towels showed a 6-fold lower impact than warm-air hand driers and a 27-fold lower impact than airblade-type hand driers [6]. However, generally lower impacts are expected from a Life Cycle Assessment (LCA) perspective. In a research carried out in 2011, the LCA procedure was applied to evaluate the overall impacts of several hand-drying methods such as virgin paper towels, 100% recycled paper towels, cotton-roll towels, standard electrical hand driers, high-speed hand driers and the airblade-type hand driers [7]. The results of the LCA of the different hand-drying methods showed that the airblade-type hand driers are expected to generate the lowest environmental impacts, followed by high-speed hand driers, cotton-roll towels, 100% recycled paper towels, standard electrical hand driers and virgin paper towels [7].

In view of such results, the present paper intends to present a methodology facilitating the decision on the convenience to install electrical hand driers in place of the conventional toilet paper towels in the restrooms of public places, based on a quantitative economic criterion and on environmental aspects. This work moves from the need for implementing sustainable actions in Italian universities that are being promoted by a recent association between Italian academic institutions [8]. The target of this study is a medium-size university located in northern Italy. At present, with a few exceptions, the large majority of restrooms in the different departments of the university are equipped with traditional paper towels for hand drying. Considered the previous results on the overall sustainability of different hand-drying methods, this paper presents a comparison between the present situation (use of paper towels) and an alternative one, based on the installation of an airblade-type hand drier in restrooms responding to a specific criterion. Indeed, installing airblade-type hand driers in all restrooms would not be a convenient solution from the economic point of view, since such hand driers present a considerable fixed cost, quantified as approximately 1,000 €. Thus, it is necessary to formulate a decision criterion to evaluate which restrooms are suitable for having paper towels replaced with electric hand driers. In addition, considerations will be expressed on the comparison between the carbon footprint generated by the present situation and the one expected in the alternative scenario. To the authors' knowledge, this is the first proposal for a methodology supporting decision processes in this field.

2 MATERIALS AND METHODS

2.1 Definition of the decision criterion

The installation of an electrical hand drier implies a fixed cost related to the purchase of the machine and variable costs related to the electric energy consumption associated with its usage. Paper towels are characterised by their purchase cost and by costs related to the collection, transportation and management of the used towels, which are regarded as residual waste. In the case of the University of Trento, the local municipality applies a waste management fare that is composed by a fixed cost and a variable cost that depends on the volume of waste generated. However, in the case of large production of waste, this cost can be approximated to the variable cost that is about 0.05 €/L. Due the presence of one scenario characterised by fixed and variable costs (airblade-type hand drier) and another scenario which entails only variable costs (paper towels), the economical convenience of installing an airblade-type hand drier in a restroom in place of paper towels can be evaluated on the basis of the average number of daily usages of the restroom by users. In other words, the target is to estimate the minimum number of daily usages above which installing an airblade-type hand drier allows for lower costs compared to the use of paper towels like in the current situation. The minimum number of daily usages is the number of daily usages that makes the annual costs of using paper towels (C_p) equal to the annual costs of an airblade-type hand drier (C_{hd}). C_p can be expressed as:

$$C_p = N \cdot n_{d,week} \cdot n_{week,y} \cdot n_{p,u}\left(C_{p,p} + C_{p,wd}\right), \quad (1)$$

where:

- N is the average number of daily usage of a restroom [-];
- $n_{d,week}$ is the number of working days in a week [d/week];
- $n_{week,y}$ is the number of working weeks in a year [weeks/y];
- $n_{p,u}$ is the number of paper towels used by each user [1/user];
- $C_{p,p}$ is the cost of each paper towel [€/towel];
- $C_{p,wd}$ is the cost for waste disposal of each paper towel [€/towel].

C_{hd} can be formulated as:

$$C_{hd} = \frac{C_{hd,f}}{T} + N \cdot n_{d,week} \cdot n_{week,y} \cdot t_{hd,u} \cdot C_e \frac{P}{1000\frac{W}{kW} \cdot 3600\frac{s}{h}}, \quad (2)$$

where:

- $C_{hd,f}$ is the initial (fixed) cost for the purchase of an airblade-type hand drier [€];
- T is the lifetime of an airblade-type hand drier [y];
- P is the power consumption of an airblade-type hand drier [W];
- $t_{hd,u}$ is the time of usage of a hand drier by a user [s/user];
- C_e is the average cost of electric energy [€/kWh].

By equalling (1) with (2), the number of daily usages of a restroom that would result in an economic benefit if paper towels were replaced with an airblade-type hand drier is:

$$N = \frac{C_{hd,f}}{T \cdot n_{d,week} \cdot n_{week,y}\left[n_{p,u}\left(C_{p,p}+C_{p,wd}\right) - \frac{P \cdot t_{hd,u} \cdot C_e}{1000\frac{W}{kW} \cdot 3600\frac{s}{h}}\right]}. \quad (3a)$$

$C_{p,p}$, $C_{hd,f}$ and P can be obtained by the suppliers, while $n_{d,week}$ and $n_{week,y}$ can be assumed as equal to 5 d/week and 50 weeks/y, considered that a university can be officially closed two weeks a year. The quantification of $C_{p,wd}$, T, $t_{hd,u}$ and $n_{p,u}$ is much more difficult, due to the high degree of uncertainty that should be considered for their estimation. Concerning $C_{p,wd}$, since the volume occupied by one package of 150 new paper towels is about 1.8 L, each new paper towel would occupy a volume of about 0.012 L. It is reasonable to assume that the volume occupied by a wet paper towel may double this value. Concerning T, the only useful information is the lasting of the guarantee issued by one supplier, which is 5 years. Regarding $t_{hd,u}$, it is reasonable to consider a mean value between 10 and 15 s. Finally, a mean value between 2 and 3 paper towels was assumed for $n_{p,u}$.

To account for the variability of the previous variables, a statistical approach was used, based on the Monte Carlo method [9]. Specifically, each variable was associated to a Gaussian probability function, characterised by a specific mean value (μ) and a standard deviation (σ). For each variable, a sample of 10,000 values was obtained, complying with the associated probability density function. Table 1 reports the values of μ and σ given to each variable characterised by uncertainty and the values of the parameters whose definition is considered as certain. N was then calculated by applying (3) to 10,000 combinations of $C_{p,wd}$, T, $t_{hd,u}$ and $n_{p,u}$.

Eqn (3) can also be rearranged as follows:

$$N = \frac{C_{hd,f}}{T \cdot n_{d,week} \cdot n_{week,y}} \cdot \frac{1}{n_{p,u}(C_{p,p}+C_{p,wd}) - \frac{P \cdot t_{hd,u} \cdot C_e}{1000 \frac{W}{kW} \cdot 3600 \frac{s}{h}}}, \tag{3b}$$

where:

- $C_{hd,f}/(T \cdot n_{d,week} \cdot n_{week,y})$ is the daily fixed cost for the purchase of an airblade-type hand drier;
- $n_{p,u}(C_{p,p} + C_{p,wd})$ is the variable cost of a single usage of the restroom equipped with paper towels ($C_{p,v}$);
- $(P \cdot t_{hd,u} \cdot C_e)/(1000 \frac{W}{kW} \cdot 3600 \frac{s}{h})$ is the variable cost of a single usage of the restroom equipped with an airblade-type hand drier ($C_{hd,v}$);

The value of N is highly affected by these values. The difference between variable costs of paper towels and of the airblade-type hand drier $C_{p,v} - C_{hd,v}$ gives the cost reduction entailed by a single use of an air drier. Notice that with $C_{p,v} < C_{hd,v}$ it is never convenient to install the airblade-type hand drier. However, since V_{air} is the energy consumption of a single usage of the air hand drier (few seconds with an hourly cost of few cents) this is an unlikely situation. With $C_{p,v} > C_{hd,v}$, which is the most likely hypothesis, the higher the difference $C_{p,v} - C_{hd,v}$, the lower is N.

Even more important in determining N is the daily fixed cost for purchase of the airblade-type hand drier. Given $C_{hd,f}$ and keeping all the rest equal, the daily fixed cost of the airblade-type hand drier heavily depends on the estimated life of the equipment, and doubling the equipment estimated life (e.g. from 5 to 10 years) halves N.

2.2 Estimation of the carbon footprint

The current use of paper towels in restrooms induces emissions of GHGs by the transportation of the product from the production facility to the final user (a university in this case) and by the transportation of waste paper towels from the user to the final waste management facility. GHG emissions from the transportation sector consists in the emissions

Table 1: μ and σ of the probability density functions assumed for Cp,wd, T, thd,u and np,u, respective minimum and maximum values obtained through the Monte Carlo simulation and values assumed for the remaining parameters.

Variables calculated via Monte Carlo simulation	μ	σ	min	max	unit
$C_{p,wd}$	0.0015	0.0002	0.0008	0.0023	€/towel
$n_{p,u}$	2.5	0.5	0.6	4.2	1/user
T	7.5	0.6	5.2	9.6	y
$t_{hd,u}$	12.5	1.0	9.0	17.9	s
Parameters assumed as certain	Reference value				
$C_{p,p}$	0.005				€/towel
$C_{hd,f}$	1,000				€
P	1600				W
$n_{d,week}$	5				d/week
$n_{week,y}$	50				weeks/y

of carbon dioxide (CO_2) both from the fuel consumption of vehicles, from the combustion of lube oil and, in the case of the latest heavy-duty vehicles, from the catalysis of urea, which is used as an additive to reduce nitrogen oxides in diesel exhaust gases [10]. In addition, minor emissions of nitrous oxide (N_2O) and methane (CH_4) occur, but such gases are characterised by high global warming potentials. They were respectively assumed as equal to 282 and 32 (i.e., as equal to the mean values of the respective ranges proposed by the Intergovernmental Panel on Climate Change, which are 265–298 and 28–36 on a 100-year time horizon [11]). To estimate GHG emissions from the transportation of paper towels, it is assumed that both the carriage of waste is carried out by heavy-duty vehicles (HDVs) with a maximum gross weight of 26 t and a tare of 10 t, in accordance with a previous work [12]. HDVs were assumed to belong to the Euro VI emissive standard [13]. Depending on the pollutant and emission mechanism, emission factors for CO_2, N_2O and CH_4 from European vehicles can be calculated or are directly proposed by a consolidated methodology developed by the European Environment Agency [10]. Specifically, According to this methodology, CO_2 emissions can be estimated as functions of the fuel consumption that depend on the mechanism of CO_2 generation. The CO_2 emission factor of a generic i-type vehicle related to fuel combustion is defined as:

$$EF_{CO_2,FC_i} = 44.011 \cdot \frac{FC_i}{12.011+1.008 \cdot r_{H:C,j}+16 \cdot r_{O:C,j}}, \tag{4}$$

where FC_i is the fuel consumption of the i-type vehicle, $r_{H:C,j}$ and $r_{O:C,j}$ are respectively the hydrogen-to-carbon and oxygen-to- carbon ratios of the j-type fuel. In the case of gasoil, $r_{H:C}$ and $r_{O:C}$ are assumed as 1.86 and 0, respectively. In the case of an HDV, FC is assumed as 240 g/km [10]. Eqn (4) applies also to the calculation of the CO_2 emissions from lube oil combustion, with the only difference that FC_i must be replaced by the oil consumption of the i-type vehicle (OC_i) and that $r_{H:C}$ and $r_{O:C}$ are respectively 2.06 and 0 in this case. OC, in the case of an HDV, is 0.156 g/km. CO_2 emissions from urea catalysis, in the present case, can be calculated as [10]:

$$EF_{CO_2,UC_i} = 0.0078 \cdot FC_i. \tag{5}$$

The N_2O emission factor for the considered HDV is 0.032 g/km, while the CH_4 emission factor for the same HDV is 0.070 g/km assuming that the transport occurs on highways [10]. The N_2O and CH_4 emission factors must be multiplied by their respective global warming potentials to give CO_2-equivalent (CO_{2eq}) emission factors. By summing up all contributions, the CO_{2eq} emission factor becomes 775 g/km. The total annual CO_{2eq} emissions essentially depend on the total travelled distance for waste carriage and on the number of travels, which can be parametrised on the basis of the annual amount of used paper towels, of the mass of each waste towel and of the maximum capacity of the HDV (16 t). Reasonably, each waste towel was assumed to be completely wet, since the transportation of waste is likely to occur before waste towels have dried. Thus, the density of waste paper towels was assumed as equal to the water density (1 kg/L). Given the assumption made on the volume of waste paper towels, the mass of each waste towel was considered as equal to 24 g.

Additional contributions are related to the final waste treatment. In the case that waste paper towels are sent to incineration, the biogenic origin of the cellulose of paper towels [14] would not generate net CO_2 emissions. In the case waste paper towels are sent to a municipal landfill, the biodegradation of cellulose contributes to the formation of biogas, whose volume fraction that is normally considered as intercepted by the biogas capture system is limited to about 75% [15]. According to a recent work [16], the CH_4 yield of cellulose during anaerobic digestion can be assumed as 0.35 Nm^3/kg of the volatile solid content, which was estimated as equal to 97.3%. Considered the molecular masses of CH_4, the global warming potential of CH_4 and the interception rate of the biogas capture system, 1 kg of waste paper towels sent to a landfill can lead to a net CO_{2eq} emission of 2.5 kg. To estimate the GHG emissions related to the landfilling of paper towels, the mass of one paper towel was measured with a laboratory scale (AS120, OHAUS Corp., USA) and resulted as 1.70 g.

The above-reported considerations allow estimating the CO_{2eq} emissions related to the use of toilet paper towels in different situations. Specifically, it is possible to express the total CO_{2eq} emissions as a function of the distance travelled by the HDVs transporting the waste towels and of the average number of daily usages of a restroom. The use of electrical hand driers entails CO_2 emissions too in the case that electric energy is produced by the combustion of fossil fuels. To estimate the CO_2 emissions from the use of hand driers, this paper considers the average grid mix of electric energy production by the different sources exploited in Italy (Table 2). By performing a weighted average of the emission factors based on the share of electric energy production by source, the average CO_2 emission factor results as 333.2 g/kWh. According to an authoritative national institute [17], this value should be increased by about 0.5% to account for CO_{2eq} emissions from the release of N_2O and CH_4 into the atmosphere, following fossil fuel combustion. Consequently, a CO_{2eq} emission factor of 334.9 g/kWh will be used to calculate the CO_{2eq} emissions induced by the use of an airblade-type hand drier ($E_{CO2eq,hd}$) according to the following equation:

$$E_{CO_{2eq},hd} = EF_{CO_{2eq},el} \cdot \frac{P \cdot t_{hd,u}}{1000\frac{W}{kW} \cdot 3600\frac{s}{h}} \cdot N \cdot n_{d,week} \cdot n_{week,y}, \quad (6)$$

where $EF_{CO2eq,el}$ is the average CO_{2eq} emission factor from electric energy production. To provide practical examples of the different impacts expected by different waste treatment strategies and different percentages of electric energy produced by renewable sources (which do not contribute to CO_{2eq} emissions), the following scenarios are proposed:

- Case 1: electric energy is entirely produced by renewable sources;
- Case 2: electric energy is produced by the average mix of sources reported in Table 2 for the case study (i.e., by 32% renewable sources);
- Case 3: waste is sent to incineration for cogeneration of heat and power;
- Case 4: waste is sent to municipal landfills.

Cases 1 and 2 will be compared to Cases 3 and 4 in Section 3.2. However, due to the uncertainties in determining the travelled distance from the transportation of paper towels

Table 2: Average grid mix of electric energy among different sources in Italy and related CO_2 emission factors [17].

Energy source		Gross electric energy production [TWh]	Share[-]	CO_2 emission factor [g/kWh]
Hydroelectric		42.8	15%	0
Thermoelectric	solid fuels	43.2	15%	899.9
	natural gas	110.9	39%	365
	derivative gas	2.2	1%	1624.8
	oil products	13.4	5%	564
	Other	21.8	8%	144
Geothermal		6.2	2%	0
Wind and photovoltaic		40.5	14%	0

from the supplier to the final user, the calculation of the total GHG emissions from the use of paper towels will be limited to the transportation of waste towels and to the final waste treatment. In addition, this study considers relatively negligible the impacts related to the production processes of hand driers and paper towels.

3 RESULTS AND DISCUSSION

3.1 Economic analysis

The results of the Monte Carlo method showed that the minimum number of daily usages of restrooms that could lead to economic benefits if paper towels were replaced by an airblade-type hand drier is between 16 and 182. This apparently wide range must be interpreted with the frequency distribution reported in Fig. 1. Such distribution is characterised by a mean N value of 36 and a mode of 32. However, a precautionary approach should be considered, especially when the investors in renovations and technology improvements are public bodies. Such an approach would suggest that a reliable trigger to decide where airblade-type hand driers should be conveniently located could be a number of average daily usages > 90. In a university, restrooms serving classrooms may respond to such characteristics.

Table 3 presents the estimated annual costs, calculated by applying (1) and (2), as a function of the number of daily usages, in the case of paper towels and of airblade-type hand driers. Costs are evaluated on the basis of minimum, maximum and most probable values of the single variable estimated through the Monte Carlo simulation.

In the present case study, 21 restrooms were identified as suitable to hosting an electric hand drier, thanks to the presence of wall sockets. Table 4 presents the estimated annual savings if one airblade-type hand drier was installed in each of the 21 restrooms. It is worth reminding that such values are only indicative, since they are influenced by the uncertainties on the variables listed in Table 1. In addition, the amortisation of the hand driers was not considered. Depending on the number of daily usages and considering the 21 restrooms, the minimum appreciable annual savings may start from a less than hundred Euro (N = 90) and

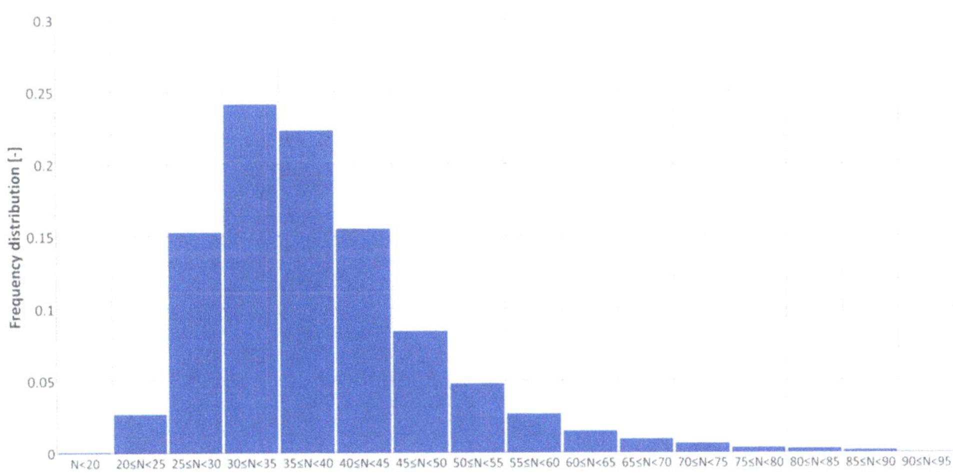

Figure 1: Frequency distribution of the minimum number of daily usages that would result in economic savings when installing an airblade-type hand drier in place of using paper towels.

Table 3: Estimated annual costs, related to one restroom, as a function of the number of daily usages, in the case of paper towels and of airblade-type hand driers.

	Annual cost of paper towels [€]				Annual cost of a hand drier [€]			
N	minimum	mean	modal	maximum	minimum	mean	modal	maximum
10	20.46	40.47	42.00	64.16	115.55	136.67	134.11	168.04
20	40.93	80.94	84.01	128.31	118.16	139.85	137.19	171.79
30	61.39	121.41	126.01	192.47	120.77	143.04	140.28	175.53
40	81.86	161.88	168.01	256.63	123.38	146.23	143.37	179.27
50	102.32	202.36	210.02	320.79	125.99	149.42	146.45	183.02
60	122.79	242.83	252.02	384.94	128.60	152.61	149.54	186.76
70	143.25	283.30	294.02	449.10	131.21	155.80	152.62	190.51
80	163.72	323.77	336.03	513.26	133.82	158.99	155.71	194.25
90	184.18	364.24	378.03	577.42	136.44	162.18	158.79	198.00
100	204.65	404.71	420.03	641.57	139.05	165.36	161.88	201.74
110	225.11	445.18	462.04	705.73	141.66	168.55	164.96	205.49
120	245.58	485.65	504.04	769.89	144.27	171.74	168.05	209.23
130	266.04	526.12	546.04	834.04	146.88	174.93	171.13	212.98
140	286.51	566.60	588.05	898.20	149.49	178.12	174.22	216.72
150	306.97	607.07	630.05	962.36	152.10	181.31	177.30	220.47
160	327.43	647.54	672.05	1,026.52	154.71	184.50	180.39	224.21
170	347.90	688.01	714.06	1,090.67	157.32	187.68	183.47	227.96
180	368.36	728.48	756.06	1,154.83	159.93	190.87	186.56	231.70
190	388.83	768.95	798.06	1,218.99	162.54	194.06	189.64	235.45
200	409.29	809.42	840.07	1,283.14	165.16	197.25	192.73	239.19

Table 4: Estimated annual savings if one airblade-type hand drier was installed in each of the 21 restrooms considered in this study.

	Estimated annual savings [€]			
N	minimum	mean	modal	maximum
10	-3,099.09	-2,020.10	-1,934.23	-1,079.21
20	-2,747.97	-1,237.16	-1,116.95	213.26
30	-2,396.85	-454.23	-299.67	1,505.73
40	-2,045.74	328.70	517.61	2,798.21
50	-1,694.62	1,111.63	1,334.90	4,090.68
60	-1,343.50	1,894.57	2,152.18	5,383.16
70	-992.38	2,677.50	2,969.46	6,675.63
80	-641.26	3,460.43	3,786.74	7,968.10
90	-290.15	4,243.36	4,604.02	9,260.58
100	60.97	5,026.30	5,421.30	10,553.05
110	412.09	5,809.23	6,238.58	11,845.52
120	763.21	6,592.16	7,055.87	13,138.00
130	1,114.33	7,375.09	7,873.15	14,430.47
140	1,465.45	8,158.03	8,690.43	15,722.94
150	1,816.56	8,940.96	9,507.71	17,015.42
160	2,167.68	9,723.89	10,324.99	18,307.89
170	2,518.80	10,506.83	11,142.27	19,600.36
180	2,869.92	11,289.76	11,959.55	20,892.84
190	3,221.04	12,072.69	12,776.83	22,185.31
200	3,572.15	12,855.62	13,594.12	23,477.78

may reach 3,570 € (N = 200). Higher savings could be reached with a higher number of usages. If considering the mean costs and the most probable costs, the estimated annual savings may start from 329–518 € (N = 40) and achieve 12,856–13,594 € (N = 200). Conversely, the installation of airblade-type hand driers would be economically inconvenient in restrooms with a low number of daily usages, as in the case of restrooms for university staff and professors. Thus, in order to pursue guaranteed savings, the installation of airblade-type hand driers is recommended in averagely-to-highly frequented public places.

3.2 Carbon footprint

The CO_{2eq} emissions related to the use of paper towels depend on two main variables: N and the distance travelled by the HDVs transporting waste towels. Therefore, the difference between the emissions generated by the scenarios involving the use of hand driers and the scenarios that consider paper towels can be expressed as functions of those two variables.

Fig. 2 presents the results of the inter-comparison between Cases 1 and 3 (best-case situations), Cases 1 and 4 (mixed situations), Cases 2 and 3 (mixed situations), and Cases 2 and 4 (worst-case situations). The results are expressed as a function of N and of the travelled distance and refer to the most probable values calculated by the Monte Carlo simulation. The results show that, if electric energy were produced by 100% renewable sources (Case 1), even in the best-case situation concerning the waste treatment of paper towels (incineration, Case 3), the use of paper towels rather than airblade-type hand driers entails GHG emissions > 0 for any N and travelled distance (Fig. 2(a)). GHG emissions would be much higher, compared to the best-case situation concerning hand driers (Case 1), if the final destination of waste paper towels were a landfill (Case 4), due to the generation of biogas and the consequent release of CH_4 into the atmosphere (Fig. 2(b)). In this case, the expected increase in GHG emissions is insensitive to the travelled distance, since the emissions from the waste treatment alternative (landfill) are higher by one order of magnitude. Both N and, to a lesser extent, the travelled distance influence the increase in the GHG emissions that would occur when preferring the use of hand driers in the current energy scenario (32% of electric energy from renewable sources, Case 2) to the use of paper towels that are sent to incineration after their use (Case 3).

In the ranges of distance and N considered here, the use of paper towels is preferable in terms of GHG emissions, if waste towels are sent to incineration (Fig. 2(c)). However, the impact of the transportation of new paper towels and of the production of hand driers and paper towels may influence such results. By comparing the two worst-case scenarios (Cases 2 and 4), the avoided GHG emissions when choosing hand driers instead of paper towels (Fig. 2(d)) are similar to those obtained when comparing the best-case scenario for hand-drier use (Case 1) with Case 4 (Fig. 2(b)). Even in this case, the increase in GHG emissions when choosing waste landfilling is insensitive to the travelled distance. This confirms the dominant role of landfilling in GHG emissions. The convenience of choosing hand driers generally increases with N also from the environmental point of view. The comparison between Cases 2 and 3 is the only exception: if a medium-low share of renewable energy is used, the environmental impacts will depend mainly on the operation of hand driers, and the impacts from the transportation of waste towels will be secondary. However, other unquantified elements may lead to different results, e.g. the GHG emissions related to the production processes of both paper towels and hand driers and the transportation of paper towels and hand driers from the supplier to the final user. By assuming similar contributions from the transportation of the two products, the production process of paper towels and hand driers may modify the results obtained. Based on a recent study [4], the environmental

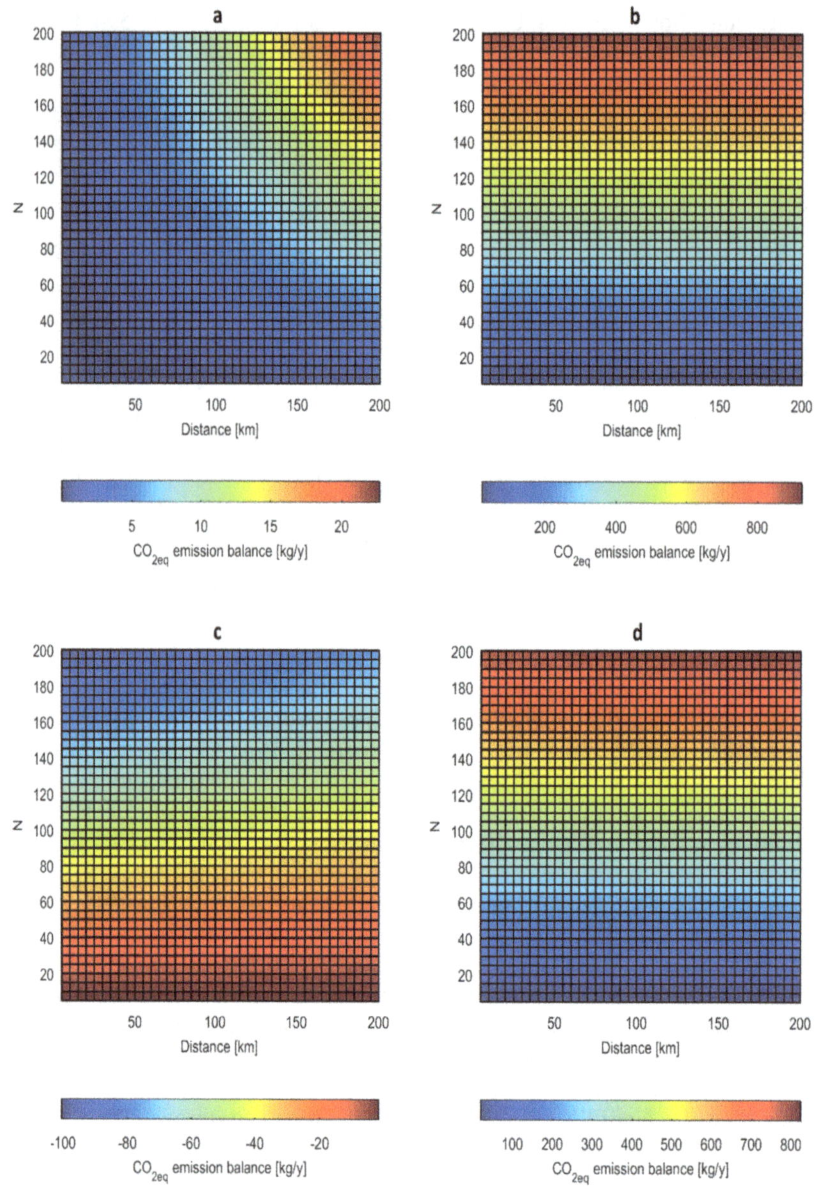

Figure 2: Differences in the CO_{2eq} emissions between (a) Cases 3 and 1; (b) Cases 4 and 1; (c) Cases 3 and 2; and (d) Cases 4 and 2, as a function of the number of daily usages of a restroom and of the distance travelled by the HDVs transporting waste towels to the waste management option; the results refer to the most probable values estimated by the Monte Carlo simulation.

impacts of hand driers are mainly related to their operation (i.e. to their energy consumption) if medium-low renewable energy is used, but GHG emissions from the manufacturing process are only slightly lower than the emissions expected from hand-drier operation. This might affect the results of the comparison between Cases 1 (100% renewable energy) and 3 (incineration), and move the convenience towards the use of paper towels in terms of GHG emissions. However, the research carried out by [7] indicates that the GHG emissions from hand-drier manufacturing are about only 11% the emissions from hand-drier operation by one order of magnitude. Thus, as emerged in a previous paper [18], LCA results cannot be generalised and require a precise definition of the context and the used data.

4 CONCLUSIONS

This paper presented a methodology to evaluate the best option between two alternatives entailing costs and environmental impacts. In this paper, airblade-type hand driers and paper towels were evaluated from the economic and environmental points of view, by identifying a criterion for assessing the economic convenience of one or the other alternative and by estimating the GHG emissions from the two. In terms of carbon footprint, the use of airblade-type hand driers would be preferable to paper towels if electric energy were produced by 100% renewable sources, independently of the waste treatment alternative. In such cases, the use of a hand drier would allow saving up to 23 kg/y and 925 kg/y of CO_{2eq} when considering incineration and landfilling, respectively. The use of an airblade-type hand drier would be preferable to paper towels if landfilling were the waste treatment alternative, even when a medium-low share of electric energy from renewable sources is available. In this case, emission savings up to 824 kg/y could be achieved, with an 89% reduction in CO_{2eq} emissions. A similar comparative scenario is considered by the LCA procedure adopted in Montalbo et al. [7]: in this case, an airblade-type hand drier would allow for a 98% reduction in GHG emissions compared to paper towels. No clear conclusions can be made when the grid mix has a medium-low share of electric energy and waste incineration is available, since information on the travelled distance for product supply and on the manufacturing processes of hand driers and paper towels would be necessary to complete the comparisons. On the other hand, the economic analysis clearly showed that the convenience of using airblade-type hand driers increases with the number of daily usages of restrooms.

When uncertainties are too high to determine the environmental impacts, economic aspects would play a greater role in determining the best option between paper towels and airblade-type hand driers. Specifically, airblade-type hand driers are preferable in restrooms characterised by a medium-high number of daily usages. On an annual basis, considering the assumptions made in Section 2.1, a restroom equipped with an airblade-type hand drier would cost 58% and 77% less than a restroom equipped with waste towels, respectively for 90 and 200 mean daily usages. Coupling economic and environmental aspects is a useful strategy to fully assess the convenience of choosing one option rather than another, without limiting the decision to only one aspect. Although the presented methodology considers the great uncertainties that characterise such an analysis, this procedure should account for additional factors that may influence the results, e.g. information on the production processes and on the logistics. Based on the available information, this study showed the potential benefits of replacing paper towels with airblade-type hand driers (in spite of their probable lower level of achievable hygiene) if an economic criterion and the different possibilities of grid mix and waste treatment are properly considered. The proposed methodology is replicable not only in Italian contexts; however, it is clear that the public adoption of hand-dryers in low-medium income countries is made difficult by factors like purchasing power and available expertise for devices maintenance.

REFERENCES

[1] Gul, M.S. & Patidar, S., Understanding the energy consumption and occupancy of a multi-purpose academic building. *Energy and Buildings*, 87, pp. 155–165, 2015. DOI: 10.1016/j.enbuild.2014.11.027.

[2] Asimakopoulos, D.A. et al., Modelling the energy demand projection of the building sector in Greece in the 21st century. *Energy and Buildings,* **49**, pp. 488–498, 2012. DOI: 10.1016/j.enbuild.2012.02.043.

[3] Jiang, P., Chen, Y., Dong, W. & Huang, B., Promoting low carbon sustainability through benchmarking the energy performance in public buildings in China. *Urban Climate,* **10**, 92–104, 2014. DOI: 10.1016/j.uclim.2014.10.004.

[4] Joseph, T., Baah, K., Jahanfar, A. & Dubey, B., A comparative life cycle assessment of conventional hand dryer and roll paper towel as hand drying methods. *Science of the Total Environment,* **515–516**, pp. 109–117, 2015. DOI: 10.1016/j.scitotenv.2015.01.112.

[5] Matthews, J.A. & Newsome, S.W.B., Hot air electric hand driers compared with paper towels for potential spread of airborne bacteria. *Journal of Hospital Infection,* **9**(1), pp. 85–88, 1987. DOI: 10.1016/0195-6701(87)90101-0.

[6] Best, E.L., Parnell, P. & Wilcox, M.H., Microbiological comparison of hand-drying methods: the potential for contamination of the environment, user, and bystander. *Journal of Hospital Infection,* **88**(4), pp. 199–206, 2014. DOI: 10.1016/j.jhin.2014.08.002.

[7] Montalbo, T., Gregory, J. & Kirchain, R., *Life Cycle Assessment of Hand Drying Systems. Technical Report.* Materials Systems Laboratory: MIT, USA, 2011.

[8] RUS—University Network for Sustainable Development, Online. https://sites.google.com/unive.it/rus/. Accessed on: 3 May 2018.

[9] Schreider, Yu.A., Buslenko, N.P., Golenko, D.I., Sobol', I.M. & Sragovich, V.G., (eds), *The Monte Carlo Method.* Pergamon Press: Oxford, pp. 1–90, 1966.

[10] European Environment Agency, EMEP/EEA air pollutant emission inventory guidebook 2016—Last Update June 2017. Technical report, 2017.

[11] Understanding Global Warming Potentials, Greenhouse Gas Emissions; U.S. Environmental Protection Agency, Climate Change Division, Online. www.epa.gov/ghgemissions/understanding-global-warming-potentials. Accessed on: 4 May 2018.

[12] Ciuta, S. et al., Role of feedstock transport in the balance of primary PM emissions in two case-studies: RMSW incineration vs. sintering plant. *UPB Scientific Bulletin Series D,* **74**, pp. 211–218, 2012.

[13] European Union, Commission Regulation (EU) No 582/2011 of 25 May 2011 implementing and amending Regulation (EC) No 595/2009 of the European Parliament and of the Council with respect to emissions from heavy duty vehicles (Euro VI) and amending Annexes I and III to Directive 2007/46/EC of the European Parliament and of the Council. Official Journal of the European Union L167/1, 2011.

[14] Lee, J., Kang, S., Kim, S., Kim, K.-H. & Jeon, E.-C., Development of municipal solid waste classification in Korea based on fossil carbon fraction. *Journal of the Air & Waste Management Association,* **65**(10), pp. 1256–1260, 2015. DOI: 10.1080/10962247.2015.1079563.

[15] Levis, J.W. & Barlaz, M.A., Is biodegradability a desirable attribute for discarded solid waste? Perspectives from a National Landfill Greenhouse gas inventory model. *Environmental Science & Technology,* **45**(13), pp. 5470–5476, 2011. DOI: 10.1021/es200721s.

[16] Wang, B., Nges, I.A., Nistor, M. & Liu, J., Determination of methane yield of cellulose using different experimental setups. *Water Science and Technology*, **70**(4), pp. 599–604, 2014. DOI: 10.2166/wst.2014.275.

[17] Istituto Superiore per la Protezione e la Ricerca Amb. (ISPRA), Emission factors of CO2 and other greenhouse gasses in the electricity sector (in Italian), Technical report, 2017.

[18] Schiavon, M., Ragazzi, M., Rada, E.C. & Merler, G., Proposal for the correct management of the Life Cycle Assessment results from integrated municipal solid waste treatment. *WIT Transactions on Ecology and the Environment* **180**, pp. 163–173, 2014.

IMPROVING THE EFFICIENCY OF THE STABILIZATION COLUMNS IN OIL AND GAS PROCESSING

DENIS ERMOLIN[1], ELENA MAGARIL[2] & ROMEN MAGARIL[3]
[1]Sosnogorsk Gas Processing Plant LLC Gazprom Recycling, Sosnogorsk, Russian Federation
[2]Department of Environmental Economics, Ural Federal University, Russian Federation
[3]Department of Oil and Gas Processing Technology, Tyumen Industrial University, Russian Federation

ABSTRACT

Rectification process is widely used in oil and gas processing and petrochemistry. The composition of the bottom product of the rectification column usually differs from the equilibrium composition due to significantly higher content of light components or fractions. This stipulates a necessity to include into the technological schemes of several oil refining processes a stabilization unit that are meant to ensure separation of gases and liquid products. A highly relevant task is to explore new possibilities for improving the stabilization process. The aim of the present work was to improve the stabilization process by changing the operating pressure. It was found that increasing the pressure in the stabilization column enhanced the sharpness of separation of butanes from pentanes, which resulted in improving the quality of the products. The experimental-industrial runs confirmed the possibility of increasing the efficiency of stabilization columns in various processes by increasing the pressure within acceptable limits.

Keywords: rectification process, sharpness of separation, stabilization column, pressure.

1 INTRODUCTION

The process of rectification has been widely used in petroleum refining and petrochemistry for quite a long time and considered by many researchers [1]–[4]. By means of rectification during the oil and gas primary processing the straight-run fractions are obtained for their subsequent use as a feedstock in the processes of production of motor fuels or in the petrochemical industry [1]. It is also important to use rectification for the stabilization of final products [5].

Due to high energy consumption in the rectification process the problems of energy efficiency of plants have attracted particular attention [6], [7]. There is an eventual necessity for improving the process technology to increase the rectification efficiency.

The choice of pressure in the rectification column is usually associated with a required temperature regime [1], [8], [9]. The studies available on the effect of pressure on the efficiency of rectification columns have often been controversial [7], [10], [11]. Previously, it was theoretically shown that the sharpness of fractionation of rectification columns in the process of operation can be substantially improved by increasing their operating pressure [12], [13]. On this basis, recommendations on ensuring the sharpness of fractionation in the deethanization columns by increasing the pressure were developed [14]. It is of considerable interest to investigate the possibility of increasing the sharpness of fractionation by changing the pressure in other rectification columns, including stabilization columns which enable separation of gases and liquid products.

2 SCOPE FOR INCREASING THE EFFICIENCY OF STABILIZATION COLUMNS BY CHANGING THE OPERATING PRESSURE

Stabilization columns are included into the technological schemes of a number of oil refining processes to improve the quality of products. The process of stabilization is the separation in stripping columns of gases, including butanes, and liquid products, starting with pentanes.

WIT Transactions on Ecology and the Environment, Vol 222, © 2019 WIT Press
www.witpress.com, ISSN 1743-3541 (on-line)
doi:10.2495/EQ180071

The process is used for the separation of gasoline from associated petroleum gas and production of gasoline in various refining processes, as well as for processing deethanized condensate.

The rate of the gas-liquid interaction is determined by the diffusion of the gas into the liquid. When increasing the pressure, the contact time between the gas and liquid phases increases and the surface tension at the gas-liquid interface decreases due to an increase in the gas density. This leads to an increase in the dispersion of gases bubbling through the liquid and effect of pressure is supposed to accelerate the diffusion interfacial mass transfer. The rate of diffusion G (the amount of substance diffused per unit time, kg/s), according to the Fick's first law, is determined by the eqn (1):

$$G = K \cdot \frac{D \cdot F}{\delta} \cdot \left(\pi \cdot y - P \cdot x\right), \tag{1}$$

where K is a coefficient, s^2/cm^2; D is the coefficient of diffusion of the gas molecules into the liquid, cm^2/s; F is the gas-liquid contact area, cm^2; δ in the thickness of the boundary film, cm; π is the overall gas pressure, Pa; P is the partial pressure of a diffusing component in the liquid, Pa; y is the mole fraction of the component in the gas vapor; x is the mole fraction of the component in the liquid.

According to the eqn (1) an increase in the overall gas pressure will lead to an increase in the diffusion rate.

3 INVESTIGATION OF THE EFFECT OF PRESSURE ON THE EFFICIENCY OF STABILIZATION COLUMNS

In order to investigate the effect of changing the pressure in the rectification system on the efficiency of the stabilization column, the experimental-industrial run was carried out on the basis of the installation for the low-temperature condensation at the Sosnogorsk gas processing plant, LLC "Gazprom pererabotka" (Sosnogorsk GPP), with a constant capacity of 280,000 m^3/h of the associated petroleum gas.

Under the experimental-industrial run of the column equipment the operating pressure range for the debutanizer was changed from 0.9 to 1.2 MPa. The upper limit of the parameter considered was limited by the maximum operating (routine) value of the installation pressure. The loading of the butane column with the column feedstock was 14 t/h.

Results of the study of the effect of pressure on the operation of the butane column are presented in Table 1 and Fig. 1.

The obtained results of the experimental-industrial run of the butane column of the Sosnogorsk GPP revealed that increasing the operating pressure in the debutanizer during the APG processing resulted in the significant improvement of the sharpness of separation of the target products.

Table 2 presents the information on the design and average statistical data on the operating mode for the stabilization columns of straight-run gasolines of several oil refineries. According to the data presented in Table 2 all the stabilization columns provided the output of the conditioned fraction IBP–180°C.

In each case, there was a margin of quality regarding the requirements for butane content.

The content of pentane in the column top product was in a range of 2.17–2.38 wt.%. The stabilization columns under consideration had comparable technological modes of operation.

The temperature regime and reflux-to-product ratio corresponded to the optimum values allowed by the technological regulations. For each plant, there was a significant scope to increase the operating pressure in the column to the values allowed by the technological

Table 1: Experimental data on the operation efficiency of the debutanizer 106-C3 of the Sosnogorsk GPP.

Parameter of the mode	Value of parameters at the given pressure, MPa				
	0.90	1.00	1.10	1.15	1.20
Top temperature, °C	60	61	63	64	66
Bottom temperature, °C	128	127	128	129	128
Data of the analytical control:					
Content of C_5H_{12} in the propane- butane fraction, %	2.16	1.11	0.37	0.10	0.05
Content of C_3H_8+C_4H_{10} in the casinghead gasoline, %	0.89	0.72	0.33	0.15	0.09

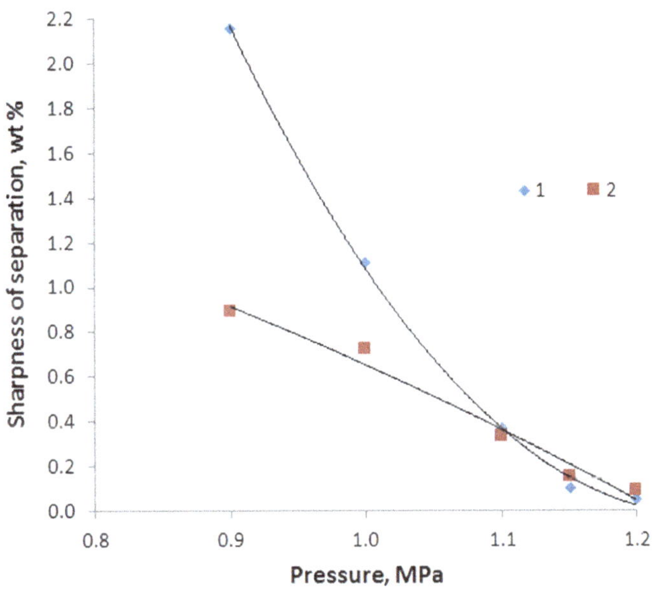

Figure 1: Influence of pressure in the butane column of the sharpness of separation: 1 – content of C5H12 in the propane-butane fraction, wt.%; 2 – content of C3H8+C4H10 in casinghead gasoline, wt.%.

regulations. All stabilization columns had the same type of contact devices – valve plates; their number for the columns under the study was in a range of 27–40 plates. The sharpness of separation in the stabilization columns, presented in Table 2 along with changes in the indicators of the technological regime and the number of contact devices, remained virtually unchanged.

Table 3 and Fig. 2 show the results of the study on the effect of pressure on the stabilization of gasoline containing 6.52 wt.% of C_4H_{10} and 18.8 wt.% of C_5H_{12}. Experimental-industrial run was carried out at the Surgut condensate stabilization plant (CSP).

Table 2: Statistical data of several petroleum refineries on the operating of the stabilization columns of straight-run gasoline fractions with boiling range IBP–180°C.

Designation of indicators	Value of indicators		
	Novoshakhtinsk refinery	LLC «Yenisei», Usinsk	Nizhnekamsk refinery
Contact devices:			
Type	Valvate		
Number, units	31	27	40
Distance, mm	700	700	700
Pressure value, MPa			
Operating	0.80	0.80	0.80
Permissible	1.40	1.80	1.40
Temperature value, °C			
Top	68	67	67
Bottom	187	190	188
Feedstock input	170	168	174
Reflux ratio	9.0	8.5	8.7
Content of C_4H_{10} in the fraction stabilized IBP–180°C, wt.%	1.63	1.89	1.47
Content of C_5H_{12} in the head fraction of stabilization, wt.%	2.38	2.24	2.17

The data obtained demonstrate the possibility of significant improvement in the efficiency of the stabilization columns of straight-run gasoline at elevated pressure, which results in increasing the sharpness of rectification. Thus, when increasing the pressure from 0.8 to 1.4 MPa (which corresponds to the allowable pressure for the stabilization columns of Novoshakhtinsk and Nizhnekamsk refineries, Table 2) in the experimental-industrial run undertaken, both C_4H_{10} content in stable gasoline and C_5H_{12} content in the head fraction of stabilization at different capacities was reduced by 57–82% and 53– 88%, respectively. When increasing the pressure from 0.8 to 1.8 MPa (corresponding to the allowable pressure for the stabilization column of oil and gas producing enterprise LLC «Yenisei», Usinsk, Table 2), C_4H_{10} content in stable gasoline and C_5H_{12} content in the head fraction of stabilization at different capacities was reduced by 86–96% and 85–97%, respectively.

Table 4 and Fig. 3 present the results of study on the effect of pressure of the process of stabilization of deethanized condensate with the density value of 677.9 kg/m^3 (at 20°C). Composition of feedstock, wt.%: C_1+C_2–0.5; C_3–12.09; i-C_4–7.14; n-C_4–10.6; i-C_5 – 5.47; n-C_5–5.29. Experimental-industrial run was carried out at the Surgut CSP.

Table 3: Operating mode of the gasoline stabilization column of the Surgut CSP at the constant output and variable pressure.

Designation of indicators	Value of indicators at the given pressure in the column, MPa					
	0.80	1.00	1.20	1.40	1.60	1.80
Feed capacity of the stabilization column of 70 m³/h						
Top temperature, °C	67	68	68	67	69	70
Bottom temperature, °C	190	190	192	190 м	192	191
Reflux ratio	8.5	8.5	8.5	8.5	8.5	8.5
Content of C_4H_{10} in stable gasoline, wt.%	1.89	1.23	0.71	0.34	0.16	0.07
Content of C_5H_{12} in the head fraction of stabilization, wt.%	2.24	1.47	0.78	0.26	0.19	0.06
Feed capacity of the stabilization column of 100 m³/h						
Top temperature, °C	67	69	68	70	70	71
Bottom temperature, °C	191	190	192	191	193	192
Reflux ratio	8.5	8.5	8.5	8.5	8.5	8.5
Content of C_4H_{10} in stable gasoline, wt.%	2.91	2.43	1.66	1.07	0.62	0.27
Content of C_5H_{12} in the head fraction of stabilization, wt.%	4.06	3.36	2.19	1.02	0.67	0.31
Feed capacity of the stabilization column of 140 м³/h						
Top temperature, °C	69	72	70	71	73	72
Bottom temperature, °C	193	195	196	194	196	195
Reflux ratio	8.5	8.5	8.5	8.5	8.5	8.5
Content of C_4H_{10} in stable gasoline, wt.%	4.99	4.47	3.31	2.13	1.39	0.69
Content of C_5H_{12} in the head fraction of stabilization, wt.%	7.14	6.20	5.01	3.37	2.08	1.10

When increasing the pressure from 0.6 to 1.6 MPa in the experimental-industrial run undertaken, C_4H_{10} content in stable condensate and C_5H_{12} content in the head fraction of stabilization at different capacity was reduced by 90–97% and 92–96%, respectively.

4 CONCLUSIONS

It was found that increasing the pressure results in increasing the sharpness of separation of butanes from pentanes, which positively affects both the balance of the process and the quality of the products obtained. Thus, in the stabilization columns, increasing the pressure

Table 4: Operating mode of the stabilization column of deethanized condensate at the constant throughput and variable pressure.

Designation of indicators	Value of indicators at the given pressure in the column, MPa					
	0.60	0.80	1.00	1.20	1.40	1.60
Feed capacity of the stabilization column of 200 m^3/h						
Top temperature, °C	74	74	76	75	77	76
Bottom temperature, °C	197	197	200	198	196	199
Reflux ratio	10.0	10.0	10.0	10.0	10.0	10.0
Content of C_4H_{10} in stable condensate, wt.%	1.40	0.84	0.43	0.16	0.06	0.04
Content of C_5H_{12} in the head fraction of stabilization, wt.%	3.08	1.87	1.12	0.54	0.31	0.21
Feed capacity of the stabilization column of 260 m^3/h						
Top temperature, °C	76	75	77	76	78	79
Bottom temperature, °C	200	197	198	199	200	198
Reflux ratio	10.0	10.0	10.0	10.0	10.0	10.0
Content of C_4H_{10} in stable condensate, wt.%	2.19	1.22	0.66	0.39	0.22	0.13
Content of C_5H_{12} in the head fraction of stabilization, wt.%	6.45	4.06	2.81	1.22	0.53	0.26
Feed capacity of the stabilization column of 300 m^3/h						
Top temperature, °C	75	77	78	76	79	80
Bottom temperature, °C	200	197	198	196	202	201
Reflux ratio	10.0	10.0	10.0	10.0	10.0	10.0
Content of C_4H_{10} in stable condensate, wt.%	3.01	2.02	1.38	0.79	0.51	0.28
Content of C_5H_{12} in the head fraction of stabilization, wt.%	9.32	6.99	4.72	3.30	1.51	0.72

within the limits acceptable for these columns can increase the efficiency of their operation. Earlier, an approach for improving the efficiency of the rectification process using surfactants was proposed [15]. It is of interest to investigate the joint application of surfactants and increasing the pressure within the acceptable limits to improve the efficiency of rectification processes.

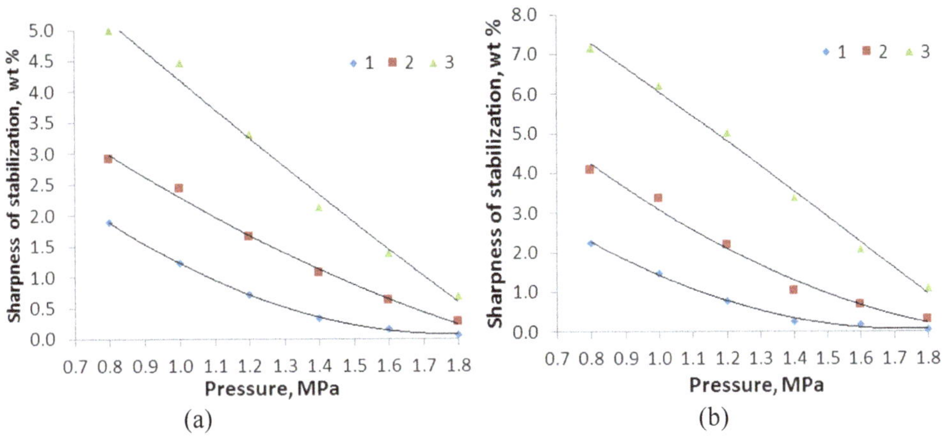

Figure 2: Effect of pressure in the column on the sharpness of stabilization of gasoline. (a) Content of C4H10 in stable gasoline, wt.%; (b) Content of C5H12 in the head fraction of stabilization wt.%. Feed capacity of the stabilization column, m^3/h: 1–70, 2–100, 3–140.

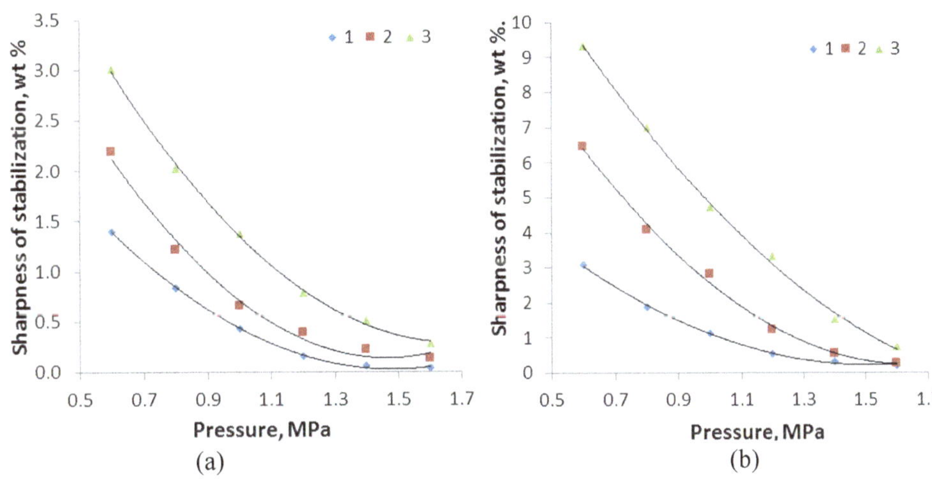

Figure 3: Effect of pressure in the column on the sharpness of stabilization of deethanized condensate. (a) Content of C4H10 in stable condensate, wt.%; (b) Content of C5H12 in the head fraction of stabilization wt.%. Feed capacity of the stabilization column, m^3/h: 1–200, 2–260, 3–300.

ACKNOWLEDGEMENT

This research was supported by Act 211 Government of the Russian Federation, contract № 02.A03.21.0006.

REFERENCES

[1] Manovyan, A.K., *Technology of Primary Processing of Oil and Natural Gas: a Textbook for Universities*, 2nd ed., Khimiya: Moscow, 2001.

[2] Scoblo, A.I., Molokanov, Yu.K., Vladimirov, A.I. & Shchelkunov, V.A., *Processes and Apparatuses of Oil and Gas Refining and Petrochemistry: A Textbook for Universities*, 3rd ed. revised and enlarged, LLC "Nedra-Business Center": Moscow, pp. 101–183, 2000.

[3] Nag, A., *Distillation & Hydrocarbon Processing Practices*, PennWell Corporation: Tulsa, pp. 1–270, 2016.

[4] Pinto, F.S., Zemp, R., Jobson, M. & Smith, R., Thermodynamic optimization of distillation columns. *Chemical Engineering Science*, **66**, pp. 2920–2934, 2011.

[5] Aleksandrov, I.A., *Distillation and Rectification in Oil Refining*, Khimiya: Moscow, 1981.

[6] Halvorsen, I. J. & Skogestad, S., Energy efficient distillation. *Journal of Natural Gas Science and Engineering*, **3**, pp. 571–580, 2011.

[7] Long, N.V.D. & Lee, M., Review of retrofitting distillation columns using thermally coupled distillation sequences and dividing wall columns to improve energy efficiency. *Journal of Chemical Engineering of Japan*, **47**(2), pp. 87–108, 2014. DOI: 10.1252/jcej.13we067.

[8] Kister, H.Z. & Doig, I.D., Studies of the effect of pressure on distillation heat requirements. *Chemical Engineering Communications*, **11**(1–3), pp. 1–12, 1981. DOI: 10.1080/00986448108910983.

[9] Jobson, M., Gorak, A. & Sorensen, E. (ed), Energy considerations in distillation. *Distillation: Fundamentals and Principles, Distillation*, vol. 1, Elsevier Academic Press: London, pp. 225–270, 2014.

[10] Liu, Z.-Y. & Jobson, M., The effect of operating pressure on distillation column throughput. *Computers & Chemical Engineering*, **23**(Supplement 1), pp. S831–S834, 1999. DOI: 10.1016/s0098-1354(99)80204-x.

[11] Castillo, F.J.L. & Dhole, V.R., Pressure analysis of the ethylene cold-end process. *Computers & Chemical Engineering*, **19**, pp. 89–94, 1995. DOI: 10.1016/0098-1354(95)87020-2.

[12] Rogalev, M.S. & Magaril, R.Z., Increasing and evaluation of the rectification tower trays efficiency, Izvestia Vyssih Ucebnyh Zavedenij. *Neft i Gaz*, **6**, pp. 92–97, 2011.

[13] Rogalev, M.S. & Magaril, R.Z., The influence of rectification sharpness on the quality of motor fuels. *WIT Transactions on Ecology and the Environment*, vol. 190, WIT Press: Southampton and Boston, pp. 833–843, 2014.

[14] Ermolin, D.B., Rogalev, M.S., Magaril, R.Z. & Trushkova, L.B., Influence of pressure on rectification sharpness in deethanization towers. *Izvestia Vyssih Ucebnyh Zavedenij. Neft i Gaz*, **5**, pp. 107–113, 2014.

[15] Ali, A.A., Magaril, R.Z., Rogalev, M.S. & Magaril E.R., Acceleration of boiling in the desired conditions. Application of its effect in refinery and petrochemical industries. *WIT Transactions on Ecology and the Environment*, vol. 186, WIT Press: Southampton and Boston, pp. 535–543, 2014.

EVALUATION OF CRUDE OILS AND NATURAL GASES OF KURDISTAN-IRAQ BY CATALYTIC IMPROVEMENTS TO LIGHTER OILS USING LOCAL CLAYS

SALAHALDIN NAMAN, LAZGIN A. JAMIL, FIRAS AL-GULAMI, SHERWAN SIMO & MAHER K. ALI
Department of Chemistry, Faculty of Science, University of Zakho, Iraq

ABSTRACT

The chemical and physical properties of three different crude oils and two natural gas wells in Kurdistan-Iraq have been studied according to American Standard Test Method (ASTM) methods. Auto-distillations have been used to determine the changes in each cut, and gas–liquid chromatography has been used to find the approximate chemical constituents. A distribution diagram of paraffin, naphthenic, and aromatic hydrocarbons has been created. The gasoline cuts for each crude oil have been examined to find suitable octane additives (oxygenated or metal organic). The octane number increases as a function of the amount of additives, which has been plotted. A correlation between the trace metals and the types of hydrocarbons has been carried out for each crude oil. The improvement of these crude oils is achieved catalytically by reactions at high temperatures using different local clays, which gave us increases in lighter molecules by 7% (max). Chemical reactions are suggested for this improvement. X-ray analysis is also performed for each clay. Research is carried out on one gas–liquid well (Khor Mor) and another pure natural gas well. Gas chromatography analysis for Khor Mor shows 51% C1–C5 and 49% C6–C11, while other natural gas wells contain 98% methane. The natural gas of the Kirkuk wells contained approximately 10–12% hydrogen sulfide, which we separated using ethanolamines, and converted to hydrogen fuel and sulfur at two pilot plants using a new catalyst instead of the previous Clause process, producing about 143 tons of hydrogen per day.

Keywords: crude oil, natural gas, evaluation, hydrogen sulfide, Kurdistan.

1 INTRODUCTION

Crude oil is a mixture of gaseous, liquid, and solid hydrocarbon compounds, and also contains small quantities of nitrogen, oxygen, and sulfur-containing compounds as well as trace amounts of metallic constituents [1]. Crude oil is usually evaluated through various physical and chemical methods. Crude oil has been known about for a long time. Archeologists have shown that it had already been extracted and used for about 5–6 thousand years BC Petroleum has also been known about for a long time. The first use of petroleum was at the Tigris–Euphrates valley, Iraq, which was inhabited as early as 4000 BC by a people known as the Sumerians who established one of the first great cultures of the civilized world [2]. Petroleum is one of the most important substances consumed by man at the present time.

Petroleum is used as a principal source of energy for industry, heating, and transportation, and it also provides the raw materials for the petrochemical plants to produce polymer, plastics, and many other products. It has also been used in very primitive ways, especially in transportation and heating, which now represent as the main source of polluted gases on the planet.

There are several theories on the formation of petroleum, which is a complex mixture of hydrocarbons in sedimentary rocks in the form of gases (natural gas), liquids (crude oil), semisolids (bitumen), or solids (wax or asphaltene) [3], [4].

In this paper, we aim to evaluate and improve the constituents of crude oil [5], natural gas [6], and the most usable cuts such as naphtha, gasoline, and diesel, so that they are less

www.witpress.com, ISSN 1743-3541 (on-line)
doi:10.2495/EQ180081

harmful to man [7], especially when they contain sulfur and poisonous trace elements [8]. Thus, we carried out intensive research to separate hydrogen sulfide from our natural gas, and then decomposed it to clean hydrogen fuel and pure sulfur [9]–[11].

2 EXPERIMENTAL ANALYSIS

All chemicals were of Analar grade and were purchased from the Fluke and Merck Co. The experimental part of the study – the extraction of hydrogen sulfide from the gas of the Kirkuk wells – was performed using ethanol amine. The decomposition of this gas to hydrogen fuel and sulfur was achieved by separating it and sending it to two pilot plants, as shown in the Figs 1 and 2 [9], [10].

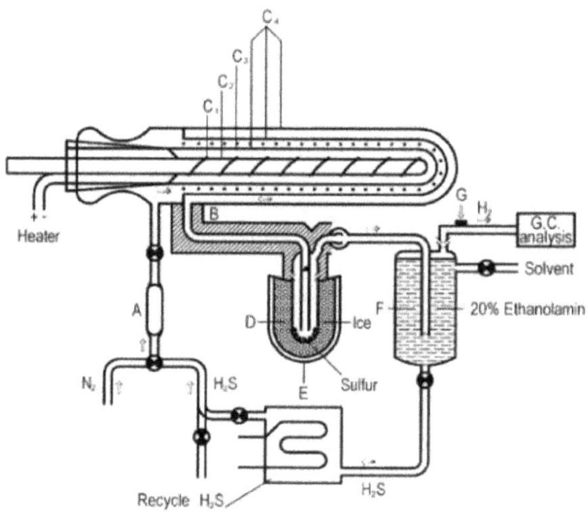

Figure 1: Pilot plant 1 for the decomposition of H_2S. A: flowmenter; B: reactor (thermal diffusion column); C1: heating wire; C2: ceramic rod; C3: catalysts; C4: quartiz; D: trap for sulfur separation; E: sulfur condenser; F: 30-wt% DEA bath for H_2S separation: G: septum [11].

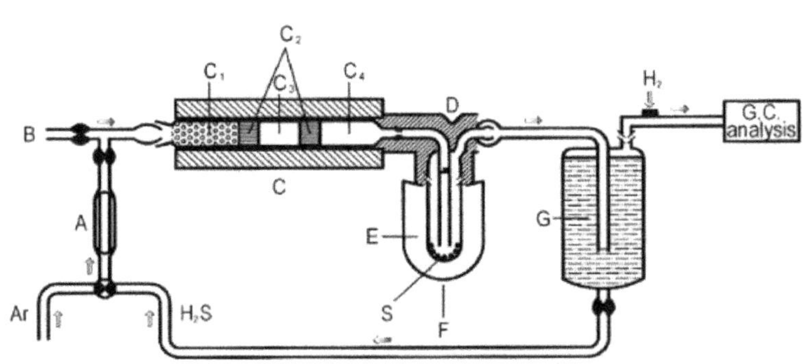

Figure 2: Pilot plant 2 for the decomposition of H_2S. A: flowmeter; B: to vacuum line; C: tube furnace; C1: pyrex balls; C2: glass wool; C3: catalysts; C4: tube reactor; D: heating tape; E: trap for sulfur separation; F: sulfur condenser; G: 30-wt% DEA bath for H_2S separation; H: gas sampling outlets septum [11].

The measurements of all physical and chemical properties have been made using modern instruments, as shown in Table 1.

Different postgraduate students have carried out systematic research work on the crude oils [5], gasoline cuts [12], diesel cuts [7], and trace metals [8] using American Standard Test Method (ASTM) and Institute of Petroleum (IP) methods for examining the physical and chemical properties of crude oils and their cuts.

3 RESULTS AND DISCUSSION (CRUDE OIL AND ITS CUTS)

The measurements of API gravity and sulfur content for three wells – Tawke, Taq-Taq, and Khurmala – are listed in Table 2.

Fig. 3 presents the distillation curves of the Tawke, Taq-Taq, and Khurmala crude oils.

A series of experiments were conducted to determine the properties of different crude oils as temperature versus accumulated volumes. Then, the carbon number of each distillate accumulated was determined using gas chromatography [5].

Tawke crude oil has a high percentage of C9, but Taq-Taq has C7, and Khurmala has C8, which represent the paraffinic hydrocarbon distributions, as shown in Fig. 4. Tawke crude oil has a high percentage of C13 and C14, but Taq-Taq has C9 and C12, and Khurmala has C9, C13, and C14, which represent paraffinic hydrocarbons.

Table 1: Apparatus and instruments employed.

1	Simple distillation apparatus ASTM D86, HAD 620 Herzog
2	Fractional distillation apparatus ASTM D 285
3	Thermometer Quickfit B14, UK
4	Gas chromatography instrument: Varian CP-3800
5	Pycnometer, (5 ml) volume, Germany
6	Viscometer, Cannon-Fenske Routine, Eurostar Scientific LTD, PSL ASTM-IP 100
7	Oxford X-ray fluorescence, PW-1600 Spectrophotometer Philips
8	Water bath, Lab Tech, model LWB-122D, Korea
9	Thermo-circulator, Lab Tech, Daihan Lab Tech Co, Model/ LCB-RO8, Korea
10	Pensky–Martens closed cup, Flash Point Apparatus
11	ACO-7 Cleveland open cup, Japan. Flash Point Apparatus
12	Muffle Furness, Lab Tech, Korea
13	Bellingham+ Stanley Ltd. Abbe 60/95 refractometer, UK
14	Shatox Sx-100 M portable octane/cetane analyzer, Russia
15	Oil product low-temperature characteristic meter OPLCM SX-800, Russia
16	Titroline KF Trace Coulometric Karl Fischer Titrator, SCHOTT Instruments GmbH, Germany
17	X-ray sulfur meter, model RX-360SH, TANAKA, Japan

Note: all of our crude oil and natural gas samples have been collected from different parts of the Kurdistan region.

Table 2: The API gravity and sulfur content of three crude oils.

Parameters	Tawke	Taq-Taq	Khurmala
API gravity	26.3	46.7	32.2
Sulfur wt%	2.82	0.65	2.48

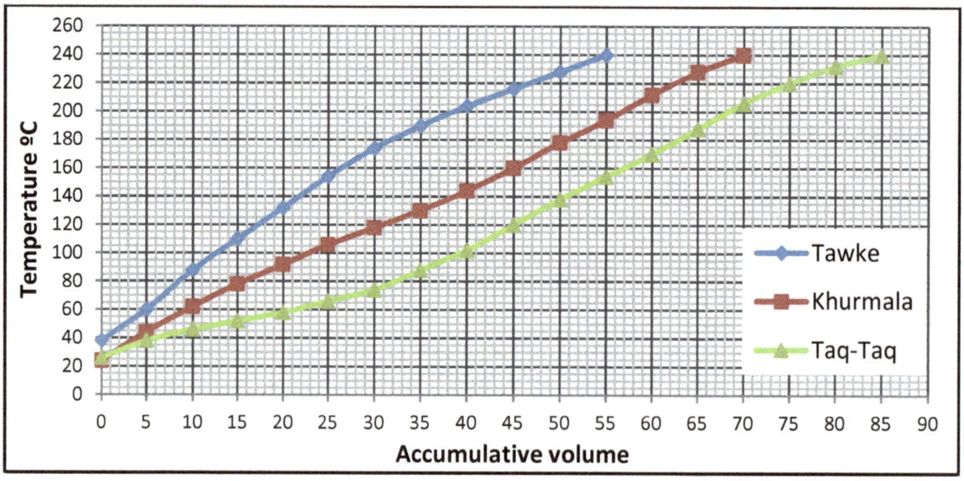

Figure 3: Distillation curve of Tawke, Khurmala, and Taq-Taq crude oils.

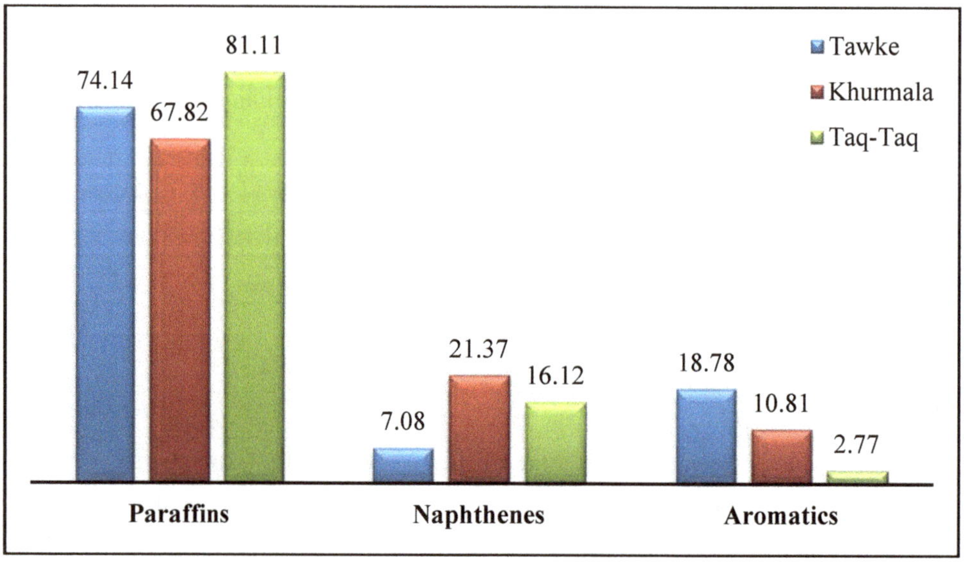

Figure 4: Paraffin, naphthene, and aromatic hydrocarbon distributions in fraction (1) at 100°C obtained via gas chromatography.

Table 3 presents the weight percentage of the hydrocarbons, starting with carbon number four of fraction (1) (initial boiling point to 100°C) and fraction (2) (from 100°C to 170°C) by chromatography, as shown in Fig. 5, as an example of the analysis of one sample.

Gasoline straight cuts are collected from the crude oils to examine the increase of the octane number by oxygenated and organometallic octane additives, shown in Tables 4 and 5 respectively [12].

Table 3: Weight percentage of the carbon number for crude oil fractions.

Carbon no. (wt%)	Light naphthafrom initial boiling point to 100°C			Heavy naphthafrom 100°C to 170°C		
	Tawke	Taq-Taq	Khurmala	Tawke	Taq-Taq	Khurmala
C4	0.00	0.34	0.42	0.00	0.00	0.00
C5	0.10	8.20	6.09	0.39	0.10	0.03
C6	1.00	22.35	16.59	1.06	0.23	0.17
C7	4.21	30.98	22.75	1.10	0.76	1.01
C8	16.76	26.69	26.79	2.24	15.29	9.07
C9	37.85	10.14	22.92	4.39	34.23	28.57
C10	16.61	0.84	2.35	3.32	14.60	9.35
C11	5.39	0.13	1.11	1.88	3.09	5.49
C12	11.05	0.20	0.48	11.94	12.00	8.36
C13	4.69	0.08	0.28	21.05	9.68	10.95
C14	1.62	0.01	0.03	21.28	5.70	10.56
C15	0.52	0.04	0.13	15.26	2.88	6.52
C16	0.20	0.02	0.06	6.72	1.44	4.43
Unknown 1	0.00	0.00	0.00	1.00	0.00	1.19
Unknown 2	0.00	0.00	0.00	2.16	0.00	2.68
Unknown 3	0.00	0.00	0.00	1.09	0.00	1.62
Unknown 4	0.00	0.00	0.00	3.36	0.00	0.00
Unknown 5	0.00	0.00	0.00	1.76	0.00	0.00
Total	**100**	**100**	**100**	**100**	**100**	**100**

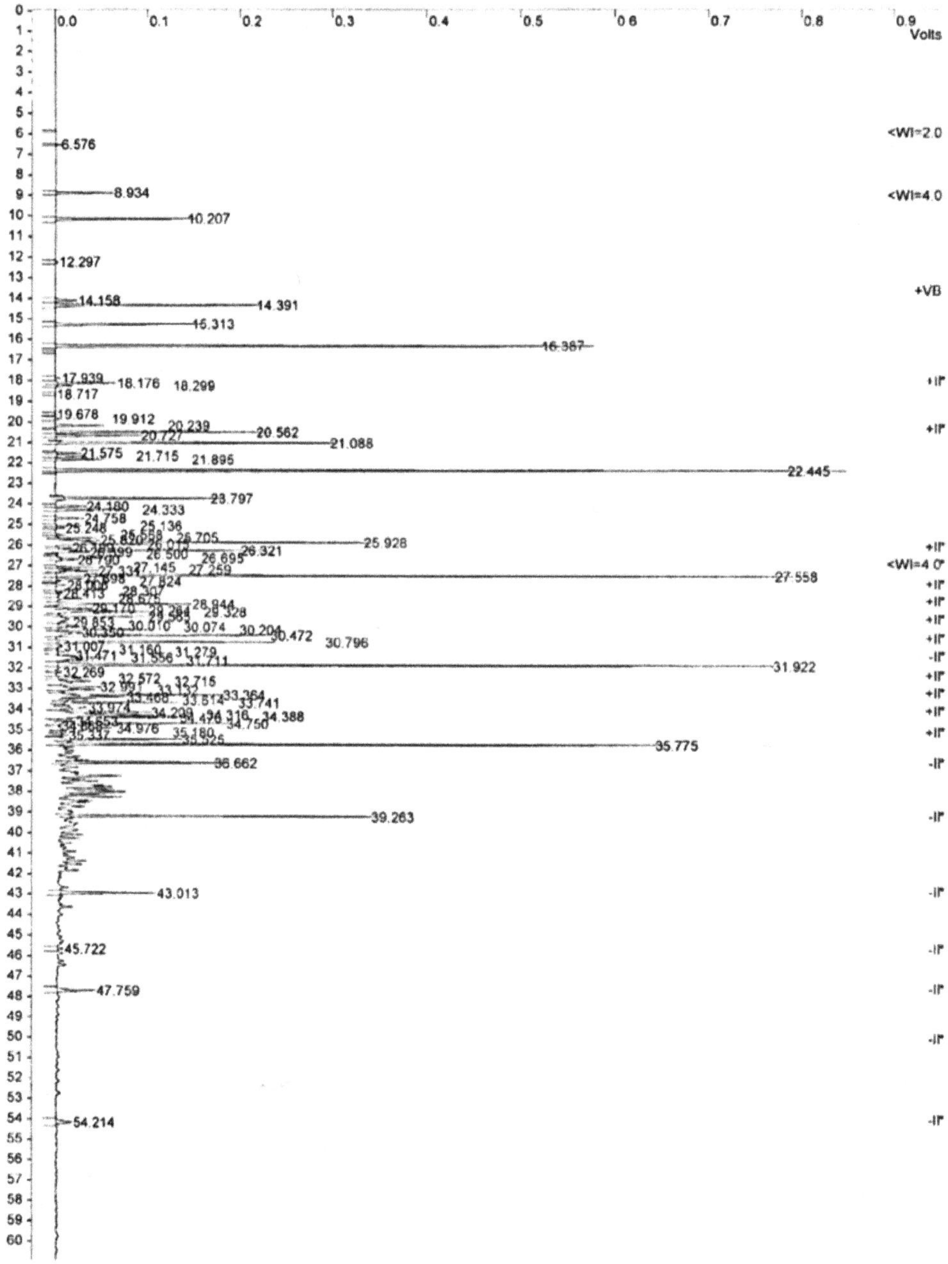

Figure 5: Gas chromatography chromatogram of Tawke crude oil with clay 3, fraction (1).

Table 4: Changes in octane number (ON) by oxygenated additives.

No.	Sample	Range of change	Δ in ON
1	TW4 gasoline	72.9–83.4	10.5
2	TW8 gasoline	72.5–83.1	10.6
3	Naphtha	71.2–88.3	17.1

Table 5: Changes in octane number (ON) by organometallic additives.

No.	Sample	Range of change	Δ in ON
1	TW4 gasoline	73.2–81.3	8.1
2	TW8 gasoline	73.0–80.0	7.0
3	Naphtha	71.8–78.4	6.6

3.1 Improvement of the crude oils with local clays

In order to improve Tawke and Taq-Taq crude oils, which represent heavy and light crude oils, three types – clays (1), (2), (3) – of clays have been collected from different places in the Duhok governorate. The results of the X-ray fluorescence analysis for the three types of local clays are illustrated in Table 6, which shows the main differences in the percentages of silica oxide, alumina, and iron oxide. The Si/(Al+Fe) ratio shows that clays (1) and (3) are similar to natural zeolite-type *mordenite* and *erionite* [5]–[7].

Fig. 6 shows that clays (1) and (3) have an ability to crack the heavy hydrocarbons in the Tawke crude oil with decreasing carbon numbers of hydrocarbon from C8 to C5 for clay (3), as well as for clay (1), but shifting to C9.

Table 6: Clay analysis.

	Clay (1)	Clay (2)	Clay (3)
Al_2O_3	7.71	6.72	6.11
SiO_2	42.05	33.37	39.42
Fe_2O_3	5.56	11.40	6.03
CaO	17.54	5.30	5.40
MgO	4.23	15.73	14.80
Na_2O	0.13	0.28	0.70
K_2O	0.96	0.43	0.95
SO_3	0.00	0.16	0.05
L.O.I	21.86	26.23	24.97
Total	100.04	99.62	98.43
Si/(Al+Fe)	3.17	1.84	3.25

3.2 Evaluation and improvement of diesel cut from Tawke crude oil wells, Zakho

For evaluating the percentage of diesel in different crude oils, a preliminary distillation at atmospheric pressure was carried out according to ASTM D86. The percent of diesel was detected for each type of crude oil and then research was carried out to increase the quality and quantity of it by taking three types of local clays [7].

According to Fig. 7, the best reflex time for increasing the volume of diesel cuts for T-8 and T-16 crude oil using 6%wt of each clay was 6 hours.

Fig. 8 shows the ability of clays (1), (2), and (3) to crack the heavy hydrocarbon in T-8 crude oil to a lighter hydrocarbon, as it is clear in the analysis of carbone distribution of diesel cut before and after.

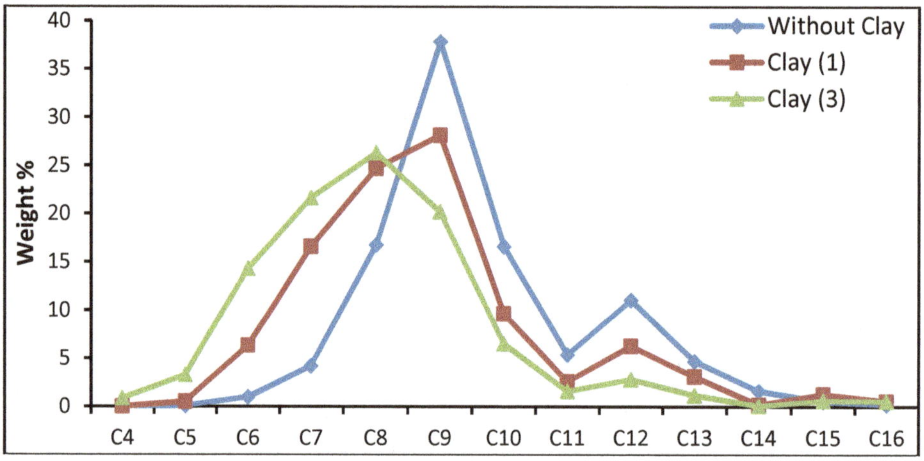

Figure 6: Carbon distribution of light fraction at 100°C before and after treatment of Tawke crude oil with clays (1) and (3).

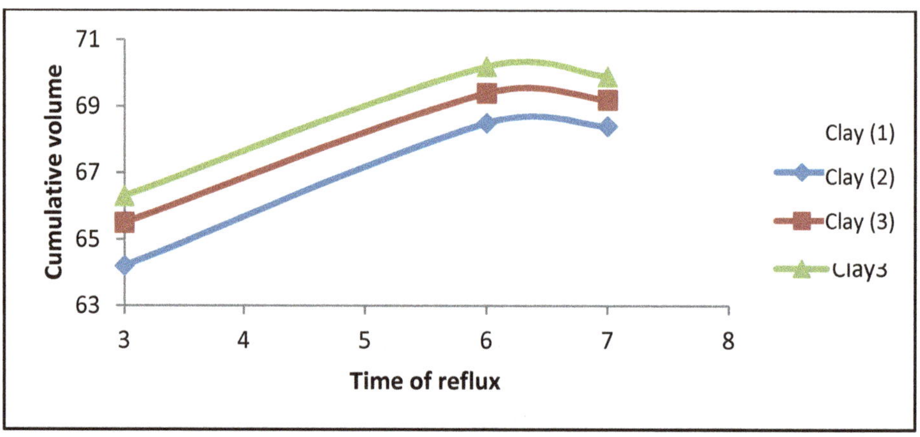

Figure 7: Effect of reflex time on the T-16 fractional distillation with 6%wt of clays (1), (2), and (3).

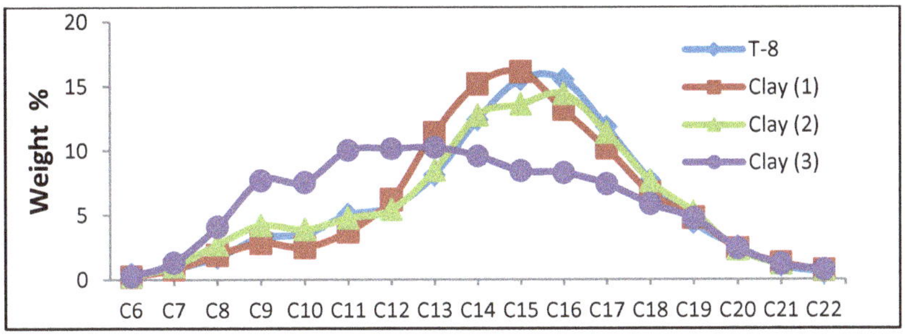

Figure 8: Carbon distribution of diesel fraction at 200–320°C before and after treatment T-8 crude oil with clays (1), (2), and (3).

3.3 Correlation between trace metals in crude oil and type of hydrocarbon

The total hydrocarbon in our sample of crude oil was 97.99%w, and its trace metals were 0.0129%w. The trace metals were determined via the carbon residue, and their ash analysis was carried out using inductive coupling plasma–optical emission spectroscopy. We also used atomic absorption spectroscopy.

The results show that this correlation depends on the characteristics of the crude oil well, including its depth, temperature, pH, and the geological formation of the wells. Most of these trace elements have a tendency to form organometallic complexes on the bottom of the wells. We did not find any relation between these trace elements and the type of hydrocarbon in the crude oil.

3.4 Separation and purification of natural gas

The distillation fraction of the gas condensate of Khor Mor has been plotted in volume percent (&v/v) as a function of distillation temperature, as shown in Fig. 9.

The gas chromatography shows the hydrocarbon-content carbon numbers C3–C13. Fig. 10 shows the percent of each carbon atom in the naphtha of Khor Mor.

The separation of C1 to C5 as a natural gas feeds two electrical power stations (Erbil and Sulamaniyh) the rest is condensate liquid (C5 to C12) as light naphtha which is transferred to the refinery to improve the octane number of gasoline.

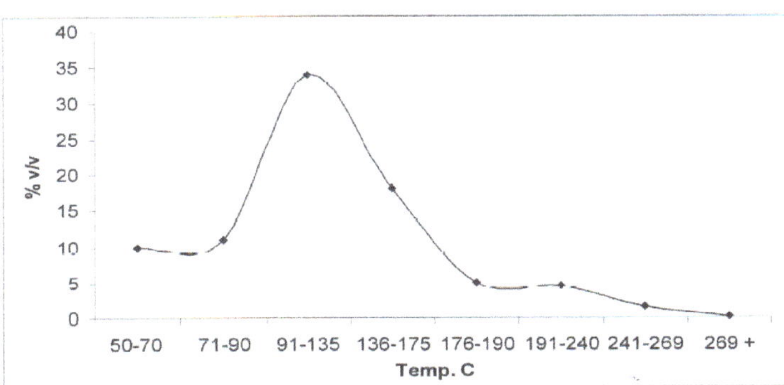

Figure 9: Fractional distillation curve.

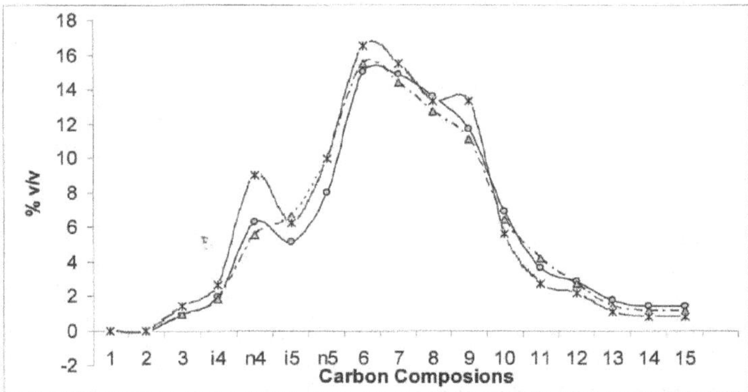

Figure 10: Analysis of the gas chromatography results.

3.5 Separation and production of hydrogen sulfide from the Kirkuk wells

Another group of postgraduate students studied the conversion of hydrogen sulfite after separating it from the natural gas of the Kirkuk wells using two different pilot plants 1 and 2, as shown in Figs 1 and 2, via thermal decomposition over a new catalyst at a temperature of approximately 400°C.

These pilot plants will convert our Clause process plants in Kirkuk to produce hydrogen gas fuel and pure sulfur, and our calculation shows that we could produce 142 tons of hydrogen per day from the natural gas of the Kirkuk wells as it contains between 10% and 11% H_2S.

4 CONCLUSIONS

This paper is a general and basic part of our research on crude oils in Kurdistan-Iraq. Further details of this work can be found in our PhD and MSc theses [5]–[8], [12]. We aim to evaluate and improve the quality of our crude oil fuels and natural gases in light of the significant polluting effects they have on human health, and as an educational lesson for the new generation about the disadvantages of classical fuel. Our target is to substitute this classical fuel with clean hydrogen fuel and solar energy in the near future.

REFERENCES

[1] James, G.S., *The Chemistry and Technology of Petroleum*, 3rd ed., CRC Press: Wyoming, 2006.

[2] Simanzhenkov, V. & Idem, R., *Crude Oil Chemistry*, CRC Press, 2003.

[3] Lur'e, M.A., Kurets, I.Z. & Shmidt, F.K., The possible abiogenic origin of oil and gas chemistry and technology of fuels and oils. *Chemistry and Technology of Fuels and Oils*, **29**(1–2), pp. 1–5, 2003.

[4] McQueen, D.R., *The Chemistry of Oil Technology by Flood Geology*, Institute For Creation Research: Santee, CA, 1986.

[5] Naman, S. & Al-Ghalami, F.S., Assessment and Evaluation Study of Crude Oils in Kurdistan. PhD thesis, University of Zakho: Iraq, 2013.

[6] Naman, S., Natural condensed gas Khor-Mor official report Gara laboratory. www.sanaman.com, 2012.

[7] Naman, S. & Simo, S., Evaluation and Improvement of Diesel Cuts, Kurdistan-Iraq. MSc thesis, University of Zakho: Iraq, 2013.
[8] Naman, S., Othman, S. & Hassein, H.M., Correlation Between Trace Metal.... MSc thesis, University of Zakho: Iraq, 2014.
[9] Naman, S., Kinetic study of thermal production of hydrogen from H_2S. *International Journal of Hydrogen Energy*, **14**(3), p. 173, 1989.
[10] Naman, S., Thermal production of hydrogen from H_2S using V_xS_y. *International Journal of Hydrogen Energy*, **15**(1), pp. 1–5, 1990.
[11] Naman, S., Pilot plant for conversion of polluted H_2S/CO_2 in the oil field to produce clean fuel of hydrogen. Presented at *Clean Energy 2000 Millennium Congress*, Geneva, Switzerland, 2000.
[12] Naman, S. & Amin, F., Evaluation and Improvement of Gasoline.... MSc thesis, University of Zakho: Iraq, 2011.
[13] Naman, S., Improvement of octane number using new nontoxic additives.... *Majalat Al-Alom*, **117**, p. 18, 2001.

REDESIGN OF AN AXIAL COMPRESSOR WITH MASS FLOW REDUCTION OF 30%

VIACHESLAV ALEKSEEVICH SEDUNIN, YURII MIRONOVICH BRODOV, OLEG VIACHESLAVOVICH KOMAROV, VITALIY LEONIDOVICH BLINOV, ALEXANDER VLADIMIROVICH SKOROKHODOV & SERGEY ALEKSANDROVICH SERKOV
Turbines and Engines Department, Ural Federal University, Russia

ABSTRACT

In this paper redesign process of an axial compressor of a Gas Turbine for mechanical drive is discussed together with computational results and experimental data. The goal of the project was to reduce compressor mass flow by 30% and at the same time to increase compressor specific work by about 10%. This could not be achieved by conventional methods such as re-staggering of Inlet Guide Vanes. Throughflow and CFD calculations were performed for redesigned versions. As a result an updated compressor was produced for the real engine and achieved design objectives. This paper shows how the swept area distribution along the compressor affects stage loading distribution and surge limits.
Keywords: axial compressor of gas turbine, CFD computations, meridional contours, flow path, profiling of compressor blades, verification of computational model.

1 INTRODUCTION

Sometimes industrial requirements can cause non-standard approaches to the engineering design. From one side, this can lead to not-the-best variant according to theoretical understanding, but from the other hand, being done in this way, it might save serious amount of funding, and therefore that non-optimality can be justified. An example of such approach is discussed in this paper.

Due to exhaust of a natural gas field at one of the customer sites a required power of a gas compressing unit dropped down by more than 30%. It is known, that thermal engine, particularly gas turbine can maintain optimal efficiency only in relatively narrow range of generated power – for best modern machines it can be 70 to 105% of optimal load, for older designs this range is less than a half of that. So, operating in the range of 50 to 70% is way below the optimum, and providing that turbine's power will not go above, it sound reasonable to perform minor corrections to the flow path of axial compressor and turbines in order to reach the efficiency peak at new operating conditions. Though, it was expected to reduce the isentropic efficiency of the compressor by about 2%. Similar re-design project was carried earlier on for better stage matching (see [1]).

As for Turbine Inlet Temperature (TIT), which can be used to compensate the efficiency drop, it was more or less limited, since the materials were already chosen for most of the parts. However, approximately 5% increase in TIT was planned based on improved combustor cooling scheme. For a turbine part the approach was similar to the compressor's one. Only compressor design is presented further in this paper.

So, the project goal was to redesign the Gas Turbine engine for reduced design power by 30%. This type of engine was already in operation for many years so minimum new parts as well as minimum re-machining were required. Also, the machines were located on site in remote area, so the optimum logistics was also considered.

As a result for axial compressor part the requirements were: i) to keep constant hub diameter of the flow path as it was in the datum design; ii) to reduce mass flow mainly by reducing the outer diameter of the flow path and blade height accordingly; iii) to leave rotor blades' stagger angles as it were, so only rotor tips can be machined down to the new

WIT Transactions on Ecology and the Environment, Vol 222, © 2019 WIT Press
www.witpress.com, ISSN 1743-3541 (on-line)
doi:10.2495/EQ180091

dimensions. As a result, an inner casing was the only major part to be manufactured, when rotor, both rotor and stator blades, outer casing and air ducts were re-used from old machines. The stator blade's configuration allowed to easily adjust them at the new casing thanks to pivot roots design.

2 PROBLEM DESCRIPTION

Aerodynamic design of an axial compressor consists of several main steps starting from one-dimensional stage matching along the flow path, then radial distribution of the flow (including blade twist along the span), and the third step deals with complex three-dimensional flow structures such as tip leakage flows, corner separations etc. Obviously, the datum compressor already accommodated successful one- and two-dimensional designs, and the 3D approach was not applicable by the time when this particular machine was developed. Therefore, the major task of the redesign project was to maintain stage matching as it was set initially but with respect to presumably higher blockage and losses due to reduced blade height.

In general, mass flow reduction could be achieved by closing the variable inlet guide vanes (VIGV). But, firstly, the compressor inlet vanes were not designed as variable. Secondly, closed VIGV leads to reduced loading of the frontal stages; however, the project goal was to increase it. And finally, the inlet guide vanes were already far too closed at the datum machine (the outflow angle is 59^0 from the frontal surface, when optimal range is 60.90^0), and further closing by 30% of the area will convert IGV to a significant throttle.

So, vanes and blades stagger angels in axial compressor were decided to keep the same as it was for datum machine, and only swept area were adjusted for lower mass flow. From practical point it is interesting to follow how the area distribution along the flow affects the stage loading and surge limits of the machine.

In present design the axial compressor has constant hub diameter with shroud contour being described by three lines, so the flow path has conical shape (Fig. 1). Stator part consists of outer and inner casings, and inner casings can be removed, re-machined and replaced. Whereas the forged rotor with cylindrical shape has tangential grooves for blade roots and is unlikely to be re-machined.

So, the most important question is the approach to mass flow reduction. The rotor diameter remains the same, so does the blade' stagger angles, whereas the tip diameter (or so-called casing diameter) is to be cut by appropriate distance. This way was chosen by customer due to its technological simplicity and more robust design, since the rotor assembly remains the same with trimmed blades and casing space plates were introduced in the non-rotating part with no need of re-staggering.

It is well known that similar kind of studies is common in centrifugal compressors where the shroud control allows to expand or move the compressor characteristic along the mass flow axis and also change its efficiency and stall margin [2]. For multistage axial compressor design several papers were also presented [3]–[5]. In [3] a rich literature overview is given.

3 MERIDIONAL SHAPE ADAPTATION

When a compressor is made of a consequence of similar stages (in literature it is called the repeating stage concept), there are two ways of matching the stages together: adjusting stagger angle of rotor and stator blades or adjusting the blade height in order to maintain constant flow angles throughout the flow path. Normally, final configuration will combine both ways.

Figure 1: Axial compressor cross-section.

The value of certain adjustments is dictated by axial speed of the core flow, which is, in turn, a function of pressure ratio of upcoming stages, effective spanwise blockage and compressor efficiency. More detailed approach can be found in [6].

So, the interesting case appear when staggering is fixed by existing blades and roots design, but the flowpath meridional profile is the only parameter to be adjusted. In order to make this possible, one should be able, first, to develop adequate model for predicting the blockage and efficiency (and, as a consequence, loading distribution along the axis). The second is to apply the model for further blade height reduction, where the combination of profile and secondary loss became crucial.

The very first iteration, which is natively expected, was to imply the proportional cut with the swept area of new compressor defined as:

$$A_{new} = A_{dat} \cdot \frac{G_{new}}{G_{dat}} \cdot \frac{\rho_{dat}}{\rho_{new}} \cdot \frac{B_{dat}}{B_{new}},$$

where A is the swept area of each stage, ρ – air density and B – effective blockage. So the mass flow ratio is corrected by change in density and blockage, where the former is a function of upcoming stages' pressure rise and efficiency, and the latter mainly depends on the endwall effects.

The end-wall losses for the 30% shorter blade will obviously be higher when the actual size of the radial gap remains the same in absolute value. But the amount of change is not that clear.

So, the problem is how to obtain and implement new pressure loss and blockage corrections in order to achieve aerodynamic similarity with the datum design. The only possible solution by that time was to have extensive CFD analysis in order to estimate compressor behavior with reduced blade height. The two most uncertain questions were: i) how good is applied CFD method in reproducing endwall flow features, and therefore losses and blockage and ii) how univocal is a correspondence between gradient in stage swept area and loading distribution? In other words, since the endwall blockage and the losses are functions of stage operating point, then could it be the case that by, say, reducing the swept area even steeper, one can reduce the blockage to the level of the datum design and, therefore, reach another equally good configuration.

These two questions are just raised here, but have no clear answer, so it was decided to do computational study in relatively wide range of swept area configurations along the flow path. On Fig. 2 different variations of shroud contours are presented as a result of different design approaches.

After couple of iterations three most suitable configurations were chosen: V2, V4 and V7. The V2 is scaled down by 25% at the inlet and 35% at the outlet with the intention to increase loading of the frontal stages, so less efficient rear stages would do less compression work. The only difference between V4 and V7 is different slope at rear stages, where V7 has simply scaled down swept area (reduced by 30%), and at V4 the last stage swept area reduced by 25% only with linear interpolation to V7 at stage 6, accounting for higher expected blockage.

4 CFD CALCULATIONS AND ANALYSIS

For CFD computations ANSYS CFX was used. Size of the element next to wall was about $y^+ \leq 100$. The total number of cells was about 2×10^6 cells. Turbulence model – k-ε standard. The interface between rows – Stage. Tip clearance dimensions were taken from "cold" geometry drawings regardless further thermal expansion of the materials. Boundary conditions were: total pressure and temperature at the inlet and static pressure at the outlet. Series of calculations were carried out with the outlet static pressure increase. Normally this approach shows good convergence at the design point and up to 10.15% above it.

Previous experience [1], [7] has shown too optimistic prediction of a mass flow comparing to the real compressor. This was explained by miss-prediction of the size of separation zones at the end-walls. However, expectations of greater mass flow proved to be wrong, and real manufactured compressor provided very close to computed value of the mass flow. Similar sort of calculations and its discussion were presented in [8], [9].

The analysis given below was performed for all variations, but for the sake of simplicity only key parameters are presented for three best variants.

Parameters at Fig. 3 are given at the design speed and also at 75% of it, which represent engine zero power conditions. On Fig. 3 the stage load distribution is presented in percentage of that for the datum compressor. For all variants the pressure ratio at the first stage (called "zero stage" due to historical process) is significantly lower, that is caused by closed inlet guide vane. This was made intentionally in order to maintain stable operation at part speed, so this approach was maintained in new machine. The rest of the stages are almost equally loaded, 7th and 8th stages at initial design are slightly overloaded. This is due to shroud

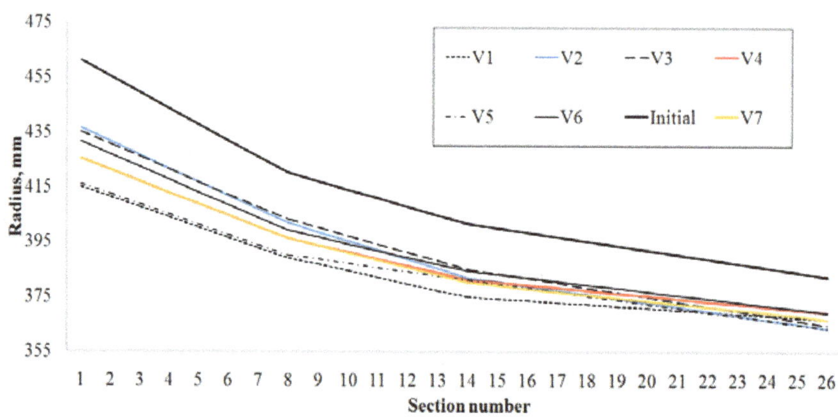

Figure 2: Shroud contours of different variations.

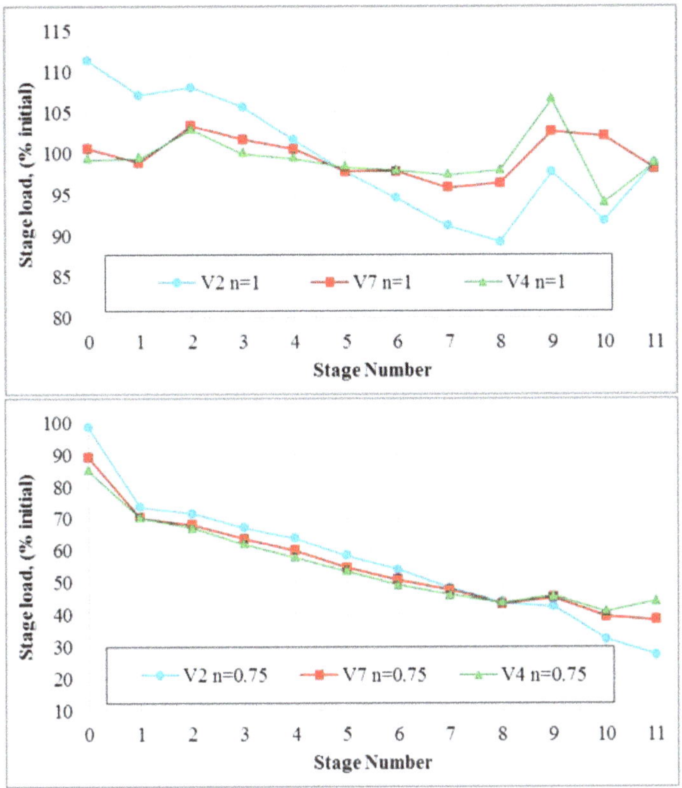

Figure 3: Load distribution along the compressor at design speed (top) and 75% (bottom).

line break after the 6th stage, which probably caused by sudden blockage change. For versions 4 and 7 the character of the lines is the same, but for version 2 the load is reducing. Loading shift in stage 9 is another example of the blockage change, since around this stage the flow separation occurs, so the effective channel area varies a lot.

So, the intention of having overly loaded frontal stages at V2 was achieved, however, for the part load, this led to unstable operation of stage one (see Fig. 3 bottom). Speed line calculations (Fig. 5) proved that V2 has lower stall margin at 75% speed, then versions 4 and 7. But due to frontal loaded design V2 has the highest compressor efficiency, as was expected initially.

On Fig. 4 Mach number distributions are presented for datum compressor, V2 and V4. For V4 and the initial compressor it can be seen the blue contours at last stages that represent more intensive flow separation. Since the flow path design is mainly two-dimensional, and the hub section is the most loaded, the flow separation means higher load at this mode. This confirms the loading distribution on the Fig. 3.

The compressor map based on series of computations is presented on Fig. 5. The upper limit was defined as the maximum outlet static pressure when the computation converged. It is known that the computation model is highly sensitive to the mesh quality and turbulence modeling, but this configuration was widely tested by authors and proved its early surge character, so the computational surge limit expected to occur at 3.5% lower pressure ratio than it would be on the real engine.

The V2 has lower surge limit at part modes, which is consistent with data shown above. Finally, V4 configuration was chosen as a base for further production and investigation due to its optimal surge limits and stage characteristics in the whole working range.

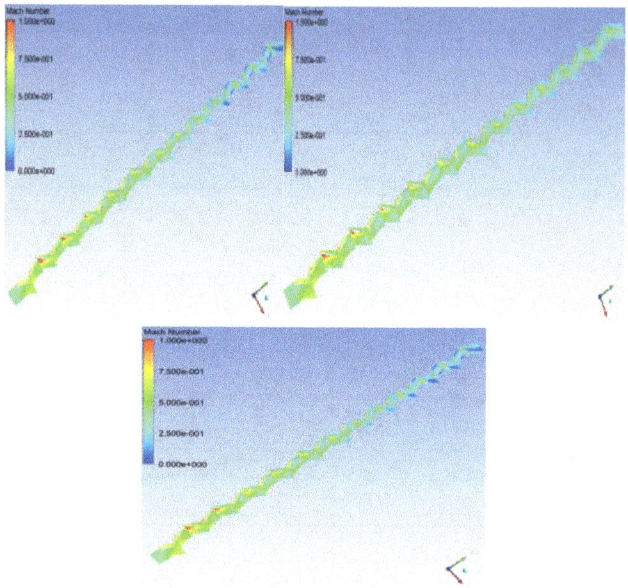

Figure 4: Mach number distributions for the datum compressor (top left), version 2 (top right) and 4 (bottom).

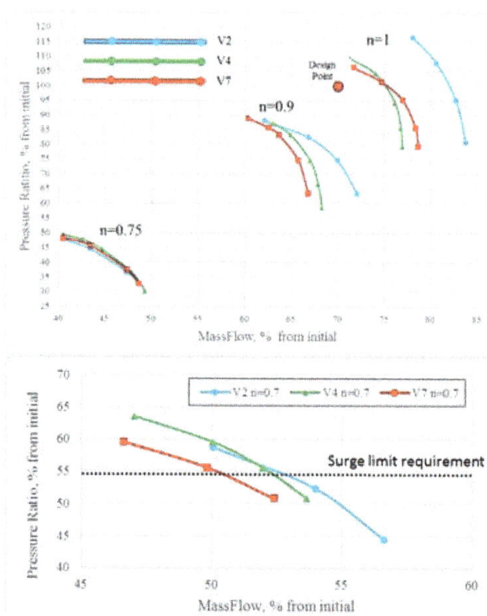

Figure 5: Compressor map for different flow path configurations

5 EXPERIMENTAL STUDY

As a result, the real engine with re-machined compressor casing, new turbine and combustor modules were manufactured and installed on site. The pre-installed instrumentation of a gas turbine allowed measurements of mass-averaged performance parameters: the mass flow at the compressor inlet, total pressure at the inlet duct and combustor outer space (essentially after compressor) and also total pressure at the cooling air channel. The cooling air is extracted between two inner casing arrangements (after the 5th stage), so, measured pressure is considered as shroud static pressure after the 5th stage rotor blade (see Fig. 1). Similar approach was used during series of tests, described, for example in [7].

The measurement set allowed estimation of turbine power, unit efficiency, and some overall performance parameters of the compressor and the turbine. Control of pressure pulsations was arranged above the second and pre-last stage of the compressor with high frequency piezoelectric sensor in order to control aerodynamic stability via pressure pulsations on part modes.

As a result, final adjustment procedure was in need due to compressor/turbine miss-matching. There were several reasons for that: lower than expected compressor adiabatic efficiency and lower than expected flow capacity of the high pressure turbine. As a result, the engine could not reach the required power at 100% speed. Also compressor stability was under concern.

In order to match compressor/turbine mass flow and consumed power two positions of compressor inlet guide vanes were tested together with two positions of first nozzle of the power turbine. Where the former were addressed to compressor air flow reduction and the latter was aimed to gain in high pressure turbine power. As a result, Fig. 6 represents the compressor adiabatic efficiency versus GT output power. The first number represents the Inlet Guide Vane position and the second is for power turbine nozzle. So for example, 0/+2 means the initial position of IGV and PT Nozzle opened by 2 degrees.

As a result, compressor efficiency went up to the expected level of 83 to 85% within the operating range. However, notable penalty in output power had to be accepted. Further work for better compressor-turbine matching is to be done after more detailed instrumentation and wider range of operating conditions to be studied. Installed piezoelectric sensors didn't show any significant pressure pulsations as possible blade resonance or rotating stall effects, so

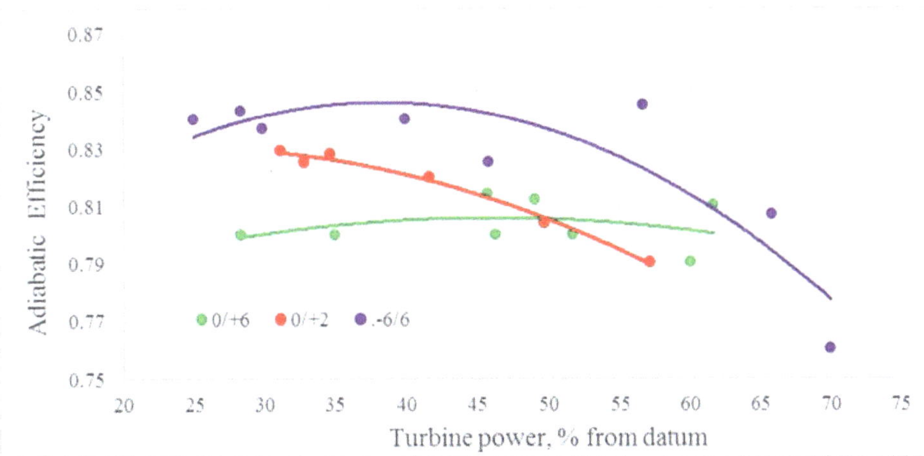

Figure 6: Compressor efficiency versus the GT power.

both first and last stages are far away from stalling. Although, the measurement set used was insufficient to judge the detailed compressor behavior, the overall performance objectives were successfully achieved.

6 CONCLUSIONS

Mathematical modeling and CFD calculations were carried out in order to choose the optimum redesign configuration for axial compressor with reduced mass flow. The shroud contour effect on stage loading distribution and blockage was studied computationally. It is shown that even 2–3% change in blade height can significantly change the stage loading and compressor stall margin.

The correction of 5% to the swept area was applied at the compressor exit in account for increased secondary losses and blockage. Maximum area correction was made at the rear group of stages, whereas for the frontal stages it was considered unnecessary.

Variations of the meridional contours showed drastic change in compressor performance, and due to insensitivity of the CFD models to many local flow features applied corrections might not be optimum in wide operating range. For the same reason fine tuning of the compressor model was skipped until further experimental data will be used for model validation.

As a result of the project, the redesigned compressor showed desired performance on site. However, more detailed instrumentation of the test compressor, such as casing static pressure measurements is planned in the nearest future for better tuning of the computational models. After final tests similar configuration is to be applied in 9 other machines.

ACKNOWLEDGEMENT

This article was prepared with the financial support of Competitiveness Enhancement Program – CEP 3.1.1.1-18.

REFERENCES

[1] Sedunin, V.A., Komarov, O.V., Blinov, V.L., Skorokhodov, A.V. & Prokopetc, A.O., The application of modern Computational fluid dynamics techniques for increasing the efficiency and stability of an axial compressor in an industrial Gas Turbine. Energy Production and Management in the 21st Century, vol. 1. *WIT Transactions on Ecology and the Environment*, vol. 190, WIT Press, 2014.

[2] Danilishin, A., Kozhukhov, Y. & Yun, V., Multi-objective optimization for impeller shroud contour, width of vane diffuser and number of blades of the centrifugal compressor stage based on the CFD calculation. *9th International Conference on Compressors and Their Systems) (City University London)*. Programme and Book Abstracts, p. 80, 2015.

[3] Ikeguchi, T. et al., Design and Development of a 14-Stage axial compressor for industrial gas turbine. *Proceedings of ASME Turbo Expo 2012*, Denmark, GT2012-68524.

[4] Wellborn, S.R. & Delaney, R.A., 2001, *"Redesign of a 12-Stage Axial-Flow Compressor Using Multistage CFD"*, ASME 2001-GT-0351, presented at the ASME Turbo Expo: New Orleans, LA, 2001.

[5] Attia, M.S., Upgrade of a 16-stage industrial compressor, Part II: Extension of the analysis method to the design function and results. *Proceedings of ASME Turbo Expo,* Barcelona: Spain, GT2006-91199, 2006.

[6] Komarov, O.V., Sedunin, V.A., Blinov, V.L. & Skorokhodov A.V., Parametrical diagnostics of gas turbine performance on side at gas pumping plants based on standard

measurements. *Proceedings of the ASME Turbo Expo: Turbine Technical Conference and Exposition*, GT 2014, Dusseldorf, Germany, 16 June 2014 through 20 June 2014—Volume 3B, GT2014–25392.

[7] Sieverding, F., Casey, M., Ribi, B. & Meyer, M., "Design of Industrial Axial Compressor Blade Sections for Optimal Range and Performance", ASME GT2003-38036. *Presented at The ASME Turbo Expo 2003*, Atlanta, GA, 2003.

[8] Larosiliere, L.M., Wood, J.R., Hathaway, M. & Dang, T.Q., *Aerodynamic Study of Advanced Multistage Axial Compressor*, NASA TP-2002-211568.

[9] Alm-Eiden, A.M., Abdel Gawad, A. F., Hafaz, G. & Abd El Kreim, M.G., Design and Optimization of a Multi-Stage Axial Flow Compressor. ICFD11-EG-4103. *Proceedings of ICFD11: Alexandria*, 2013.

CRITICALITIES AND POTENTIALITIES OF LOCAL RENEWABLE SOURCES OF ENERGY

LUCA ADAMI[1], GIACOMO CASTAGNA[1], ELENA MAGARIL[2],
RAMONA GIUREA[3], NAVARRO FERRONATO[3], GIANLUCA RUGGIERI[3],
VINCENZO TORRETTA[3] & ELENA CRISTINA RADA[1,3]
[1]Department of Civil, Environmental and Mechanical Engineering, University of Trento, Italy
[2]Ural Federal University, Department of Environmental Economics, Ekaterinburg, Russia
[3]Insubria University, Department of Theoretical and Applied Science, Varese, Italy

ABSTRACT

In this article, we will discuss about the potential of renewable energies available in a relatively small territorial context, such as municipalities, and the various problems interfering with a full exploitation of said potential. Since the launch of Covenant of Mayors, CoM, in 2008, thousands of local governments committed on volunteer basis in reaching or even exceeding, to the EU targets with regards to climate change and renewable energies for the years to come. Every local government committed to the CoM needs to produce the SEAP (Sustainable Energy Action Plan), which highlights the actions that the government is willing to implement in order to achieve the target set by the CoM (referred either to the 2020 or to the 2030). While studying the local context, many problems of various nature often arise which can be detrimental for the full exploitation of the potential of renewable resources available. This article displays a few examples of local governments located around the world and the problems encountered while writing their SEAP, related to geographical position, types of agricultural, industrial and commercial activities in the area, plants already operative, financial situation of the local government and the population living in the area. The possible areas of intervention include energy efficiency works on public and private buildings, traffic reduction strategies, energy recovery systems applied to factories and industrial plants, solar thermal, photovoltaic and wind power installations, material and energy recovery processes applied to waste management and hydroelectric power installations.

Keywords: availability, planning, renewable energy, sustainability, targets.

1 INTRODUCTION

Paris agreement on climate change sets a long-term temperature goal of holding the global average temperature increase below 2°C, possibly limiting the increase to 1.5°C above pre-industrial levels [1]–[4]. Presently, Paris agreement has been ratified by 178 of 197 Parties at the Convention [5]. In order to achieve these goals, several climate-and energy-related initiatives at different levels has been introduced in the last decade [6]–[9]. The initiative called Covenant of Mayors (CoM) for Climate and Energy was launched in 2008 with the aim to gather local governments to achieving the EU climate and energy targets. On a voluntary base, municipalities or groups of them (called "signatories" by the CoM initiative) can adhere to CoM and sign a declaration stating that they plan to obtain specific targets of CO_2 emissions reduction through specific local actions that are described in the SEAP. In particular CoM signatories endorse a shared vision for the long-term target of 2050: accelerating the decarbonisation of their territories, strengthening their capacity to adapt to unavoidable climate change impacts, and allowing their citizens to access secure, sustainable and affordable energy [10]. A short-term target consists in actions to support implementation of the EU 40% greenhouse gas-reduction target by 2030 and the adoption of a joint approach to tackling mitigation and adaptation to climate change. CoM initiative is part of the world climate change networks such as the Conference of Parties 21, "C40 cities", the ICLEI – Local Governments for Sustainability and the Smart Cities Initiative.

www.witpress.com, ISSN 1743-3541 (on-line)
doi:10.2495/EQ180101

CoM initiative also monitors the progresses of the signatories through two types of Progress Plans: The Monitoring Report, related to the mitigation and adaptation actions adopted, that has to be compiled every two years and the Emission Inventory, related to resilience to climate change, reduction of emission and energy consumption, that has to be compiled every four years. Progress plans consider seven sectors in which specific targets are identified: industry, local electricity production, local heat cold production, municipal buildings equipment facilities, public lighting, residential buildings, tertiary buildings, transport [11]. In the decade 2008–2017, 5960 SEAPs and 1741 progress plans have been published with 2020 targets. The coverage of progress plans is currently below 30%. Only best practices have been traced so far, while local problems SEAPs adoption haven't been described in detail.

In this paper we present 5 cases covering 4 continents describing the geographic characteristics, the economic situation and the origin of the energy locally consumed for each case. Moreover, we focus on the aspects of the environmental sensibility of the population and on a general description of the public environmental programs and of the case studies related to renewable energies in order to point out potentialities and criticalities given by the local renewable sources of renewable energy of the case studies.

2 CASE STUDIES

2.1 La Paz (Bolivia)

Geographic characteristics. La Paz is located in central Bolivia, near the Andean plateau and the Cordillera. It has a high rate of urbanization (about 1.1%) [12] and poor political urban planning. The city has a population of about 900,000 inhabitants and it is the seat of the National Government. For this reason, it could be considered one of the most developed Bolivian cities. Its average altitude is about 3,600 m above sea level. It is located within valleys and rifts, with different slopes, which do not allow the introduction of new infrastructures and factories. Local urban planning management issues are enhancing the need of water and energy, representing a real barrier for improving the sustainability of the city.

Economic situation. La Paz is inserted into a low-middle income country, with about 3,070 USD of gross national yearly income (GNI) per capita and with about 39% of the population below the national poverty line [13]. Therefore, the differences between rich and poor people are always more visible. The city is one of the most developed within the country due to a tertiary sector well developed. However, the economic sustainability of the city is mainly supported by the national government, in particular regarding the environmental management, which is not covered by municipal taxes, and the costs of the fuels, which are reduced for guaranteeing a low cost of the transportation.

Origin of the energy locally consumed. The support at national level of the fossil fuels is mainly due to amounts produced at national level. Indeed, Bolivia is a producer of methane and petrol, which is used both within the country and exported to neighbour countries, such as Peru, Paraguay, Argentina and Brazil. For this reason, the electricity generated in Bolivia derived mainly by fossil fuels (60%), of which 91% is produced from methane combustion, and for 40% by hydroelectric power [14]. Therefore, the use of renewable energy is not well developed and encouraged.

Environmental sensibility among the population. As a developing big city, the public awareness about environmental issues is low, due to the lack of public campaigns and the introduction of environmental activities. The lack of financial sustainability could not encourage the improvement of the environment, supported only by international NGOs.

Public environmental programs and case studies related to renewable sources of energy. There are not local and national programs related to the use of renewable energies, as well as significant pilot projects related to renewable energy sources. The application of new technologies for the exploitation of other source of energy are only applied in pilot projects, or household level. It is the case of the anaerobic digestion, for the production of bio-methane for cooking, or the application of solar panels for the generation of hot water. However, these technologies are not encouraged at national level. The opportunities and weak points regarding renewable energy sources are reported in Table 1.

2.2 Ekaterinburg (Russia)

Geographic characteristics: with a population of 1,48 million inhabitants, Ekaterinburg is located in the eastern side of the Ural Mountains. The city is located on a natural watershed, with the presence of several water bodies in the neighbourhood of the city. There are two lakes in the city. Ekaterinburg's position on the border of Europe and Asia offers opportunities as a centre of cooperation between East and West.

Economic situation. Ekaterinburg is a historical and economic centre of the Urals forming a natural border between Europe and Asia [15]. The city is very rich in minerals and raw materials; main industry branches are heavy machinery and metallurgy. An increasing number of international companies operate their branches and joint ventures here. The city is the administrative, economic and cultural capital of the Urals. Nowadays the city is an important centre of industry, transport, finance, science and culture [16]. Ekaterinburg is one of the largest financial and business centres in Russia. Key industries present are: ferrous and non-ferrous metals, machine engineering and metal processing, energy [17].

Origin of the energy locally consumed. The city's heat and power are supplied by five power stations. However, overall consumption of energy is between 30–40% higher than in comparable Western European cities. This is coupled with high levels of atmospheric pollution [18]. Electrical energy feeding the city is produced by power plants working on natural gas. Shortfalls are mainly covered by nuclear energy delivered from a power station [19]. Based on available data of the population and total energy demands in the region, energy annual consumption per capita was estimated as 10.5 MWh [20]. A methodology was proposed for including the environment as one of the factors in the regional electrification programs [21].

Environmental sensibility among the population. An emerging attention to renewable energies can be linked with the city, in particular through its candidacy as the hosting city of EXPO-2025 event, on the theme: “Changing the World: Innovations and Better Life for Future”. City will develop in the next decade several actions on the topics of energy saving, energy efficiency, power industries including hydro-electric power plants.

Public environmental programs and case studies related to renewable sources of energy. The city is presently investing funds in the development of the Multi-Year Investment Plan for Municipal Environmental Infrastructure, started in 2003 [22] and in specific studies on the quality of water from sources of decentralized water supply [23], use of wood in the “Green Project” to enhance the green endowment of the city [24] and as an energy resource [25]. Renewable energy sources use for sustainable development of the city was also analysed [26] as well as fees as a source for the environmental development of the city [27]. The increased quality of life is related to the environmental development of the city [28].

Table 1: Critical issues in the potential exploitation of renewable sources in La Paz.

Source type	Potential exploitation	Criticalities / Opportunities
Livestock waste	The city and the areas around are not a source of significant livestock waste flows. So, no specific waste treatment plant are implemented.	-
Sewage sludge	No wastewater treatment is present (even decentralised) thus no sewage sludge is generated	The expected development of a strategy for wastewater treatment in the city will change the scenario towards the generation of sewage sludge.
Organic fraction of municipal solid waste (OFMSW) and green waste	The quantities of waste generated within the city are significant, although not treated; they are disposed of into a sanitary landfill.	There are many management issues regarding OFMSW management. Only a few pilot projects are introduced for the treatment of such streams. Moreover, the application of anaerobic digestion is very difficult at an environmental and technological point of view.
Biogas from landfill	The sanitary landfill is a source of biogas which is not exploited. Moreover, is production is not monitored and well managed.	The landfill biogas could be considered an option for the generation of electric energy. However, it is not encouraged, and the investment required could represent a barrier in this context.
Surface water bodies	Hydropower plants are introduced in the country. However, the city is mainly provided by electric energy generated by fossil fuels.	The morphology of the city could be exploited for the implementation of hydropower plants. However, specific investments are required, although the policy of the country is mainly focused on the exploitation of fossil fuels.
Photovoltaic	It is not a technology well developed in Bolivia, while La Paz could not be considered as the best area for its application due to the lack of space for the building of photovoltaic parks.	The particular characteristics of the area of La Paz make necessary the development of a specific photovoltaic panel. However, it could be applied at household level, and the solar incidence could be advantageous for its application.
Solar thermal	This technology is implemented at household level. However, it is not introduced within the city, since its main application could be found in rural areas and within boundaries of the city.	This technology could be of interest for the area, although it could not be considered as the main solution for solving the energy demand.
Wood (local)	Wood and other biomasses, such as agricultural waste, are mainly used in the external areas of the city and in the rural areas. The centre of La Paz is not organized for its use.	The combustion of biomass is a big issue as regard indoor pollution in rural areas, while in La Paz is not applied. This could be considered an advantage in an environmental point of view, since the combustion of methane, instead of biomass, allows reducing the generation of NO_2 and PM_{10}.

2.3 Sibiu (Romania)

Geographic characteristics. Sibiu County is a county (județ) of Romania, in the historical region of Transylvania, with the capital city Sibiu. The county of Sibiu is situated in the centre of Romania, at 280 km away from Bucharest, the capital of the country. It is a county of middle size, with a total area of 5.433 km^2 [29], [30]. The city is in the temperate continental zone, with thermal influences due to the depression area and the mountains that surround it in the south and south-west [31].

Economic situation. The main economic branches with a high share in the economy of the municipality are: production of components for the automotive industry, logistics and logistics packaging, food industry, textiles, services and production craft. Sibiu is an important economic hub for Romania, with a high rate of foreign investments [32]. These activities add to the services that support the cultural tourism sector, based on the existing heritage on the cultural agenda and the reputation of the European Capital of Culture in 2007. From the perspective of the dynamics of economic activity, the financial crisis has affected the number of firms; however, the unemployment rate remains low [33]. In the Forbes Best Cities top 2015 – 40 best business cities, Sibiu ranks 13th. Sibiu is already ranked 5th at national level in terms of GDP per capita and 6th in terms of net average wage. The industrial profile is given of the food industry, the textile apparel industry, wood processing, building materials and the automotive industry (car parts and subassemblies). Tourism is also an important sector, based on natural and cultural potential the region. A factor that plays an important role in the economy of the city is tourism, which has been increasing at a steady rate since 2007. Sibiu is one of the most important cultural and economic centres of Romania and was designated the European Capital of Culture for the year 2007, along with the city of Luxembourg [34]. Formerly the centre of the Transylvanian Saxons, the old city of Sibiu was ranked as "Europe's 8th-most idyllic place to live" by Forbes in 2008 [35]. In 2019 Sibiu County will be awarded the title of Region of Gastronomy, alongside Greece's South Aegean region [36]. The factors making the economic development of the county are the geographic position, natural resources and handicraft tradition in processing the county's resources.

Origin of the energy locally consumed. *1. Thermal energy supply*: In the city of Sibiu, natural gas is used to provide thermal energy. Heating of buildings is done either in a centralized system or using individual boilers or stoves. At the level of the city of Sibiu, natural gas distribution is provided by S.C. E-On. Gaz Distribuţie România S.A. Under the Sustainable Energy Action Plan of the City of Sibiu, 2014 the gas distribution network totals 420 km, the total number of branching at the level of the City of Sibiu being 19,802. The technical state of the gas distribution network is good in 90%, the remaining 10% being rehabilitated. [37] *2. Electricity supply*: On the territory of Sibiu Municipality, the electricity distribution to the consumers is ensured by the Sibiu Electricity Distribution Branch belonging to S.C. Electrica SA - Transylvania South Branch. Supply of electricity on the regulated market is mainly done by S.C. Electrica S.A., as the default supplier. At the local level, the power supply is made by means of five transformer stations, connected to the National Power Grid, with a total capacity of 190kV. As far as street lighting is concerned, it is carried out with 7,852 sodium vapour lamps with nominal power between 70W and 250W respectively with an installed power of 1,356.28 kW. The analysis of lighting sources carried out for the elaboration of the Sustainable Energy Action Plan of the City of Sibiu in 2014 [37] reveals that about 42% of them have a small wear degree, the remainder being of high and medium wear. Thus, the replacement of high-wear sodium-vapour lamps is a consideration with high-performance luminaires with low consumption electricity. The city of Sibiu is supplied with electricity from the National Power System through 5 transformer power stations.

Environmental sensibility among the population. In the current context, taking into account climate change and global warming, nature is receiving ever greater attention. From the primary sector (agriculture) to the secondary sector (industry, production) or tertiary (services), human activities extract their resources from nature. Thus, it is essential that mid-term and long-term planning take account of the environmental impact. Being responsible, in the current understanding, refers to aspects of lifestyle, sustainable development, and being "green". This attitude presupposes the desire of members of the community to make a positive contribution, aware of the impact that human settlements have on the environment. A responsible company uses the resources that it has at its disposal, without creating any disadvantages for future generations [38].

Public environmental programs and case studies related to renewable sources of energy. Romania is Member States of the European Union, therefore EU law affects direct or indirect the national [39], [40] and regional law [41]–[44] and becomes part of the legal system. The European Union is a source of law for Sibiu city. In Sibiu the actions are focused on the transpose of European regulations in internal legislation and on the assurance of implementation conditions to European requests. Furthermore, in Romania the legislative background was improved by adopting the lows regarding the renewable energy chapter.

2.4 Oppeano (Italy)

Geographic characteristics. The Municipality of Oppeano is set in the southern part of the Province of Verona, in northern Italy. In this Municipality live 9,972 inhabitants, scattered among the center of the Municipality and the various hamlets (Cà degli Oppi, Vallese, Mazzantica, Villafontanta and Ponte Ebreo) in an area 49.95 km^2 wide. The municipal territory is crossed by three creeks: Bussè, Piganzo and Menago. The climate of Oppeano can be described as Cfa according to the Köppen–Geiger climate classification system; in Italy the Municipality is identified with the climatic zone E and has an HDD value of 2,340.

Economic situation. This Municipality has a very developed agricultural sector and its most iconic product is *Vialone Nano Veronese*, a renowned variety of rice. Very developed is also the breeding sector, with numerous piggeries, poultry and cow farms scattered around the territory. The presence of the industrial and service sectors is strong as well, with a multinational clothing company that have a warehouse in the municipal territory and the recently built national logistics centre of a discount supermarket chain. Potential renewable energy resources available in the Municipality of Oppeano are listed in Table 2. The scenario described in Table 2 is very promising at first look: Municipality of Oppeano has plenty of potential renewable energy resources coming from numerous production activities and environmental plants. However, with further study and insights of the situation, the possibilities of implementing a strategy to produce renewable energy starting from the resources present in the municipal area turned out to be worse than expected. The problems arose and the difficulties encountered while planning the municipal strategy for achieving the CoM 2020 emission targets are shown in Table 3. Table 3 describes the difficulties encountered to set up an efficient management strategy for the exploitation of renewable energy resources with the aim to achieve emission reduction targets.

Origin of the energy locally consumed. *1. Thermal energy supply:* in the Municipality of Oppeano are used 9.04 m^3/inh. of natural gas per year for the heating of the public buildings, and the trend shows that the quantity is decreasing (15.1 m3/inh. in year 2011). For what concerns the private buildings sector, in Oppeano are used 4,150,362 m^3/year of

Table 2: Potential renewable energy resources available.

Resource type	Quantity	Description
Zootechnical slurry	Poultry: 365,192	-
	Pigs: 8,520	-
	Cows: 6,999	-
Sewage sludge	226 kgTDS/d	Referred to the soon to be built sewage treatment plant, designed for 8,000 PE.
OFMSW	1,012 t/year	The available quantity raises to 6,500 t/year when considering the whole competence area of the local waste manager, which includes nearby municipalities. The overall quality of the OFMSW collected is high.
Green Waste	637 t/year	-
Other biomasses	Not specified	In the Municipality there are various types of crops: corn, orchards, sunflower, rice, tobacco and soy. The scraps coming from the agricultural sector could be used to produce green energy, for example in anaerobic co-digestion plants.
PV	15 public buildings	Annual global radiation on horizontal surface is 5.090 MJ/m^2. When considering also private buildings, such as houses and industrial sheds, the quantity can increase up to three orders of magnitude.
Solar thermal	15 public buildings	
Surface water bodies	3 hydraulic jumps	Located at Villafontana (Menago river), Feniletto (Irrigation channel), Villa Anti (Piganzo river).
Geothermal	-	The Municipality area is of no interest with regards to these types of energy, according to the provincial energy management plans.
Eolic	-	

natural gas, 232,932 l/year of gas oil, 338,237 l/year of LPG and 1,780,620 kg/year of biomass (wood). In Oppeano there are no district heating networks nor geothermal plants active. *2. Electricity supply:* the electricity consumption for public streets illumination in the Municipality of Oppeano is 937,757 kWh/year and the trend is decreasing (1,221,517 kWh/year in 2011).

The local government is currently financing the complete substitution of the traditional sodium-vapour lamps with LED lamps. The electricity consumption of public and private buildings is 11.884.125 kWh/year. Currently in the Municipality there are 186 PV plants active, accounting for a total power of 6.300,09 kW. The local government is evaluating to implement more PV plants on public buildings covers. In Oppeano there are no active winds farms nor hydroelectric plants.

Environmental sensibility among the population. Currently most of the Italian population is really concerned about environmental thematics, and terms like sustainability, green economy, climate change and renewables are ever recurrent in the political, scientific and public debate. With regards to this specific case, it is useful to look at the data available for the energetic certification of the residential buildings in Oppeano [45], which demonstrate the progressive increment of residential buildings certified with higher energetic classes. Moreover, with regards to the transportation sector, it should be noticed that the number of electric and hybrid cars in Veneto region is increasing year by year (6470 hybrid cars and 213 electric cars in 2015 and 12535 hybrid cars and 408 electric cars in 2017) thus confirming again the importance of environmental topics among the population.

Table 3: Critical issues encountered in the exploitation of available renewable resources for the municipality of Oppeano (Italy).

Resource type	Hypothesized exploitation	Issues
Zootechnical slurry	Anaerobic digestion to produce biogas/biomethane.	In the Municipality there is a recently built nitro-denitro plant designed for the pre-treatment of zootechnical slurry that needs to be feed. This economically prevents the use of the resource for the biogas/biomethane production.
Sewage sludge		The Municipality recently approved the project of the new sewage treatment plant, which is designed to operate in prolonged aeration conditions. The resulting sewage sludge is therefore not suitable as input in an anaerobic digestion process.
OFMSW and green waste		The availability of OFMSW in the municipality is not sufficient to feed a traditional anaerobic digestion plant under economically favorable conditions, even when considering the total OFMSW collected by ESA-com, the local waste management company.
Surface water bodies	Installation of hydraulic turbines to produce hydroelectric energy.	The jump located at Villafontana is the most interesting with 1 m of height and an average flow of 1÷2 m^3/s. It could host micro turbines of about 20÷30 kW. However, the Reclamation Consortium of Verona expressed a negative opinion about its exploitation because of hydrogeological instability of the area and the resulting costs to be incurred for additional maintenance. The other two hydraulic jumps identified in the Municipality area are too small to be exploited under economically favourable conditions.
PV	Installation of PV panels on the roofs of municipal buildings to produce electric energy.	No issues observed.
	PV noise barrier installed on the SS 434 Transpolesana road (867 kW).	This PV plant is not working now. There is a legal dispute in progress caused by the bankruptcy of the company that installed it.
Solar thermal	Installation of solar thermal panels on the roofs of municipal buildings to produce hot water.	No issues observed. However, it should be noted that for a public building the optimal solar exploitation is PV, since schools and city halls are uninhabited for the most part of the day and they need less hot water than the typical residential building does.
Other biomasses	Anaerobic co-digestion to produce biogas/biomethane.	The problems related to the scale factor mentioned above for the OFMSW stay relevant in this case too.

Public environmental programs and case studies related to renewable sources of energy. An unconventional solution to this problem, which partially address the necessity of the Municipality to reduce CO_2 emissions without compromising the operation of the plants already present, revolves around the thermochemical hydrolysis process, hereby briefly described. The OFMSW provided by ESA-com (6,500 t/year) should be used to run an anaerobic digestion plant based on the thermochemical hydrolysis process.

This process includes a pre-treatment of the OFMSW that separate the high COD, readily biodegradable liquid fraction from the low COD, slowly biodegradable semi-solid fraction. The liquid fraction is digested in a small anaerobic reactor with lower HRT than traditional ones, while the semi-solid fraction can be sent to composting with no further treatment.

This solution is compatible with the small quantity of MSW available because: (I) the digested material can be treated in the already active nitro-denitro plant with no further costs (agreement between the Municipality and the farm); (II) the exhausted air coming from the anaerobic digestion plant can be sent to the activated sludge tank of the municipal sewage treatment plant, saving here also on the treatment costs and (III) the semi-solid fraction of the OFMSW can be sent to composting at a lower cost than normal, since a part of the COD has been extracted during the thermochemical hydrolysis process.

2.5 Tanzania

Geographic characteristics. The local administration in Tanzania is based on a multilevel organization. In rural Tanzania the village council elects a chairperson that represent the village in the Ward. The Ward is a democratic subdivision that includes different villages (or different streets in urban territories). A District includes different Wards. District councils coordinate the activities of the township authorities and village councils.

Economic situation. The village and township councils have financial responsibility over the funds they receive from the national government. Mainland Tanzania (i.e. excluding Zanzibar) has 138 urban districts including 630 wards and 133 rural district which include 2682 villages. The estimated population is around 50 million in 2017. The village and township councils should also formulate plans for their areas. Plans are developed in association with formally established bodies and must secure district approval. The crucial actor in planning activities is the Ward Development Committee (WDC). The WDC coordinates development plans and social service plans, supervises project implementation and service delivery activities, and is an intermediary for discussing initiatives arising from the sub-ward levels and development plans from the higher tier local authorities.

Origin of the energy locally consumed. Tanzania's energy supply depends mainly on biomass. Since 85–90% of the population is not connected to the electricity grid, the overwhelming majority of households use wood and charcoal for cooking. As a total, biomass makes up close to 90% of the total primary energy consumption in Tanzania. Unfortunately, this leads to the deforestation of 100,000 h per year, of which only about a quarter is reforested. Other energy sources are petroleum, which makes up 7.8% of total primary energy consumption, natural gas (2.4%), hydropower (1.2%) and coal/peat (0.3%). About 6.6 percent of primary energy needs to be imported, primarily from Uganda (8 MW) and Zambia (5 MW) (International Energy Agency).

Environmental sensibility among the population. The Tanzanian government has joined the Sustainable Energy for All Initiative (SEforALL) launched in 2011 by the United Nation Secretary-General Ban Ki Moon. In 2015, the Ministry of Energy and Minerals has published an Action Agenda [46] that defines the following national targets for 2030 (compared with 2012 as a baseline year): (1) At least 75% of population with electricity access; (2) At least 75% of population with access to modern cooking solutions; (3) Rate of improvement in energy intensity at least -2.6% per year; (4) At least 50% of electricity and 10% of heat from renewable energy sources (including off-grid generation). The Action Agenda lists all the existing plans active in Tanzania in the fields of Energy Access, Renewable Energy and Energy Efficiency. Seventeen plans address the issue of electricity connections, five plans address the issue of grid development, four plans work on the clean cooking or cook-stoves issue. There are at least six plans or strategies in the field of renewable energies and at least seven plans or strategies in the field of energy efficiency.

Public environmental programs and case studies related to renewable sources of energy. Given this rather complex framework, it is possible to highlight that the main barrier to develop effective and comprehensive planning and define adequate funding is the weak communication mechanism between local government authorities and the national government. The activities on energy planning, implementation and monitoring are developed with limited or no intervention from regional authorities and other local stakeholders. The lack of communication is a big problem under different points of view: (I) the central government risks to fail to understand the situation and needs at regional level; (II) the strategies developed may have better success when including and empowering final beneficiaries and (III) if we put the specific targets in perspective of the 2030 Agenda, energy is a possible tool to the full empowerment of women: a single approach may benefit Sustainable Development Goals targets 7 and 5.

3 CONCLUSIONS AND RECOMMENDATIONS

Uttermost weather, increased temperatures, melting glaciers are obvious signs that climate change is happening. Therefore, even in a relatively small territorial context, such as municipalities, towns and regions play an important role in the actions against global warming - one of the significant challenges nowadays- taking into consideration their task regarding spatial planning, energy and transport.

As of today, thousands of local and regional governments have joined the Covenant of Mayors. In this context, this paper presents the status of a few examples of governments located around the world and the problems that are coming across while writing their SEAP, related to geographic characteristics, economic situation, origin of the energy locally consumed, environmental sensibility among the population and some public environmental programs and case studies related to renewable sources of energy. All of these are presented with the aim of underlining the potentialities and criticalities elaborated by the local sources of renewable energy encountered in those case studies.

ACKNOWLEDGEMENT

This research was supported by the Municipality of Oppeano (Italy) and by Act 211 Government of the Russian Federation, contract № 02.A03.21.0006.

REFERENCES

[1] Rogelj, J. et al., Paris Agreement climate proposals need a boost to keep warming well below 2°C. *Nature*, **534**(7609), pp. 631–639, 2016. DOI: 10.1038/nature18307.

[2] Schleussner, C.-F. et al., Differential climate impacts for policy-relevant limits to global warming: The case of 1.5°C and 2°C. *Earth System Dynamics*, **7**(2), pp. 327–351, 2016. DOI: 10.5194/esd-7-327-2016.

[3] Schleussner, C.-F. et al., Science and policy characteristics of the Paris Agreement temperature goal. *Nature Climate Change*, **6**(9), pp. 827–835, 2016. DOI: 10.1038/nclimate3096.

[4] Hulme, M., 1.5°C and climate research after the Paris Agreement. *Nature Climate Change*, **6**(3), pp. 222–224, 2016. DOI: 10.1038/nclimate2939.
[5] United Nations Climate Change, Paris Agreement: Status of Ratification. https://unfccc.int/process/the-paris-agreement/status-of-ratification. Accessed on: 30 Mar. 2018.
[6] Lombardi, M., Pazienza, P. & Rana, R., The EU environmental-energy policy for urban areas: The Covenant of Mayors, the ELENA program and the role of ESCos. *Energy Policy,* **93**, pp. 33–40, 2016. DOI: 10.1016/j.enpol.2016.02.040.
[7] Cosmi, C., Salvia, M., Di Leo, S., Pietrapertosa, F. & Loperte, S., Interregional cooperation as a key tool for the achievement of strategic-energy and climate targets: The experience of the interreg IVC RENERGY and SEE RE-SEEties Projects. *Smart and Sustainable Planning for Cities and Regions. SSPCR 2015. Green Energy and Technology,* eds A. Bisello, D. Vettorato, R. Stephens & P. Elisei, pp. 335–352, 2017.
[8] Gargiulo, M. et al., An integrated planning framework for the development of sustainable and resilient cities: the case of the InSMART project. *Procedia Engineering,* pp. 444–453, 2017.
[9] Bertoldi, P., Kona, A., Rivas, S. & Dallemand, J.F., Towards a global comprehensive and transparent framework for cities and local governments enabling an effective contribution to the Paris climate agreement. *Current Opinion in Environmental Sustainability*, **30**, pp. 67–74, 2018. DOI: 10.1016/j.cosust.2018.03.009.
[10] Covenant of Majors, Signatories' vision. www.covenantofmayors.eu /about/covenant-initiative/objectives-and-scope.html. Accessed on: 30 Mar. 2018.
[11] Delponte, I., Pittaluga, I. & Schenone, C., Monitoring and evaluation of sustainable energy action plan: Practice and perspective. *Energy Policy*, **100**, pp. 9–17, 2017. DOI: 10.1016/j.enpol.2016.10.003.
[12] INE (2012) Características de Población y Vivienda, Censo Nacional de Población y Vivienda, Bolivia. 10. (In Spanish). www.indec.gob.ar/ftp/cuadros/poblacion/censo2010_tomo1.pdf. Accessed on: Jun. 2018.
[13] World Bank Bolivia, The World Bank data. https://data.worldbank.org/country/bolivia. Accessed on: Jun. 2018.
[14] Ministerio de Hidrocarburos y Energía, Memoria Institutional. Estado Plurinacional de Bolivia (In Spanish). https://es.scribd.com/document/240107124/Memoria-Institucional-2012. Accessed on: Jun. 2018.
[15] Nifantova, R.V., Makarova, M.N. & Kosmin, I.F. Ekaterinburg on the demographic map of Russia: Facts and prospects. *Economy of Region*, **4**, pp. 52–59, 2010. DOI: 10.17059/2010-4-5.
[16] Economics of Russian cities and urban agglomeration, Institute for Urban Economics. www.urbaneconomics.ru/centr-obshchestvennyh-svyazey/news/reyting-stolichnyh-gorodov-rossii-ot-fonda-institut-ekonomiki#_ftn1. Accessed on: 12 Mar. 2018.
[17] Russian Union of Exibition and Fairs, RUEF Exibition Centres. http://ruef.ru/userfiles/ExhibitionCentresEn2016.pdf. Accessed on: 15 Mar. 2018.
[18] European Commission, TACIS Energy Centre opened in Ekaterinburg. https://cordis.europa.eu/news/rcn/4100_en.html. Accessed on: 30 Mar. 2018.
[19] Karpova, A. & Tredler, I., *Ekaterinburg: A Power Efficient City*, Siemens: Moscow, pp. 12–18, 2010.
[20] http://sverdl.gks.ru/wps/wcm/connect/rosstat_ts/sverdl/ru/statistics/sverdlStat/. Accessed on: 5 Mar. 2018.

[21] Gitelman, L.D., Gitelman, L.M. & Kozhevnikov, M.V., Factoring environment into electrification management in a region. *International Journal of Sustainable Development and Planning,* **13**(4), pp 707–717, 2018.

[22] Multi-Year Investment Plan for Municipal Environmental Infrastructure in Ekaterinburg, Russia. www.oecd.org/russia/multi-yearinvestmentplanformunicipal environmentalinfrastructureinekaterinburgrussia.htm. Accessed on: 30 Mar. 2018.

[23] Konshina, L.G., The assessment of the quality of water from sources of decentralized water supply of Ekaterinburg and surrounding areas. *Gigiena I sanitariia*, **95**(5), pp. 413–416, 2016.

[24] Damary, R., Pryadilina, N., Zalesov, S. & Opletaev, A., Outdoor wooden furniture as a component of an overall municipal project: "Green city", *28th International Conference on Wood Science and Technology, ICWST 2017: Implementation of Wood Science in Woodworking Sector*, 133147, pp. 237–242, 2017.

[25] Khudyakova, G.I., Danilova, D.A. & Khasanov, R.R., The use of urban wood waste as an energy resource. *IOP Conference Series. Earth and Environmental Science*, **72**(1), p. 012026, 2017. DOI: 10.1088/1755-1315/72/1/012026.

[26] Volkov, A. & Aristova, A., Opportunity and prospect analysis of RES utilization for sustainable development of Ekaterinburg city in Russia. *IOP Conference Series: Earth and Environmental Science,* **72**(1), 012024, 2017. DOI: 10.1088/1755-1315/72/1/012024.

[27] Bystrova, T. & Larionova, V., Fee development as a tool for the integrated formation of the urban environment, international multidisciplinary scientific geoconference surveying geology and mining ecology management. *SGEM*, **3**, pp. 239–246, 2016.

[28] Lavrikova, Y., Akberdina, V. & Mezentseva, E., Strategic guidelines of a megalopolis's development: New industrialization and ecological tension. *IOP Conference Series: Earth and Environmental Science,* **72**(1), 012011, 2017. DOI: 10.1088/1755-1315/72/1/012011.

[29] Sibiu Municipality. www.sibiu.ro/index.php/sibiu/prezentare. Accessed on: 1 Apr. 2018.

[30] Sibiu County Tourism Association. www.sibiu-turism.ro/About-us.aspx. Accessed on: 1 Apr. 2018.

[31] Ogimet CLIMAT summary for 15260: Sibiu (Romania)—Section 2: Monthly Normals. CLIMAT monthly weather summaries. Accessed on: 25 Jun. 2017.

[32] Chamber of Commerce, Industry and Agriculture of Sibiu. www.cciasb.ro/. Accessed on: 25 Jun. 2017.

[33] Strategia Integrată de Dezvoltare Urbană Durabilă a Municipiului Sibiu 2023. www.sibiu.ro/new/pdf/SIDU%20Sibiu%202017%2005%2022_small.pdf. 2017.

[34] Sibiu Cultural Capital Website. http://sibiu2007.ro. Accessed on: 1 May 2018.

[35] Beckett, E. & Olson, P., In Pictures: Europe's Most Idyllic Places to Live, Forbes. www.forbes.com/2008/11/18/europe-homes-dollar-forbeslife-cx_po_1118realestate_ slide_4.html. Accessed on: 13 Feb. 2018.

[36] International Institute of Gastronomy, Culture, Arts and Tourism (IGCAT), European Region of Gastronomy. www.europeanregionofgastronomy.org /platform/sibiu-2019/. Accessed on: 13 Feb. 2018.

[37] SE1: Sustainable Energy Action Plan of the City of Sibiu, 2014. www.sibiu.ro/ro2/ paed/memoriu_anexe_PAED-SIBIU_ONLINE_final.pdf. Accessed on: 13 Feb. 2018.

[38] SE3: Strategia de Adaptare la Schimbările Climatice a Municipiului Sibiu. www.sibiu.ro/new/pdf/SASC+_PASC_Sibiu_final_(1).pdf. Accessed on: 13 Feb. 2018.

[39] Indexul Societăţii Durabile ISD-România-2008. www.anpm.ro/anpm_resources/migrated_content/uploads/9590_3918_Index%20National%20SD%20Romania%202008.pdf. Accessed on: 13 Feb. 2018.

[40] Strategia Naţională pentru Dezvoltare Durabilă a României Orizonturi 2013-2020-2030 (National strategy for sustainable development of Romania horizons). www.anpm.ro/anpm_resources/migrated_content/uploads/9347_sndd.pdf. Accessed on: 13 Feb. 2018.

[41] Strategia Integrată de Dezvoltare Urbană Durabilă a Municipiului, Sibiu 2023 (Integrated Strategy for Urban Development of the Municipality), Publication year 2017. www.sibiu.ro/new/pdf/SIDU%20Sibiu%202017%2005%2022_small.pdf. Accessed on: 13 Feb. 2018.

[42] Planul de Acţiune pentru Energie Durabilă Municipiului Sibiu, 2014 (Action Plan for Sustainable Energy). www.sibiu.ro/ro2/paed/memoriu_anexe_PAED-SIBIU_ONLINE_final.pdf. Accessed on: 13 Feb. 2018.

[43] Strategia privind Adaptarea la Schimbările Climatice în Municipiul Sibiu (Strategy for the adaptation of climatic changes in the Municipality). www.sibiu.ro/new/pdf/SASC+_PASC_Sibiu_final_(1).pdf. Accessed on: 13 Feb. 2018.

[44] PROGRAMUL OPERAŢIONAL REGIONAL, POR Axa prioritară 3. http://regio-adrcentru.ro/axa-prioritara-3-1-sprijinirea-eficientei-energetice-gestionarii-inteligente-energiei-si-utilizarii-energiei-din-surse-regenerabile-infrastructurile-publice-inclusiv-cladirile-publice/. Accessed on: 13 Feb. 2018.

[45] Energy levels of buildings in Oppeano. https://venet-energia-edifici.regione.veneto.it/VeNet/statistiche.php. Accessed on: 13 Feb. 2018.

[46] Tanzania's SeforALL Action Agenda, December 2015. www.seforall.org/sites/default/files/TANZANIA_AA-Final.pdf. Accessed on: 30 Mar. 2018.

A NEW BIPV BRICK PAVING THE WAY TO ENERGY-GENERATING CONCRETE FAÇADE

MATTHIEU HORGNIES, ISABELLE DUBOIS-BRUGGER & FABIENNE LEGRAND
LafargeHolcim R&D, 38291 St Quentin-Fallavier, France

ABSTRACT

A significant part of the world global energy consumption is used to heat and/or and cool buildings. While insulation regulations are gradually strengthened in several counties, decentralized energy generation (such as the building-integrated photovoltaic, BIPV) is more and more considered as a strong lever to improve the energy self-consumption of the building and mitigate the impact of the large urban areas on climate change. Our researches aim to promote new photovoltaic systems on the concrete-based envelop (more especially by the way of an aesthetical cladding system) in order to increase the value of residential and office buildings. A durable multi-layer's system based on ultra-high performance fibre-reinforced concrete (UHPFRC), specific structural adhesives and solar cells (or films) has been developed to be directly cladded on the façades of buildings. After demonstrating the durability of the multilayers using different aging tests in laboratory, the production of a fully integrated cladding system, based on concrete, has been demonstrated in pre-cast plant in 2017. The next step, under progress, is the building of an outdoor demonstrator, which will be helpful to give proofs of durability under weathering and monitor the efficiency the PV façade panels for different sun exposures over several years.

Keywords: BIPV, adhesion, durability, photovoltaic, concrete, interface, façade, demonstrator.

1 INTRODUCTION

A part of the global energy consumption is used for heating and cooling buildings and insulation regulations are gradually strengthened all over the world. Furthermore, decentralized energy generation in buildings is considered as one of the main levers to mitigate the impact of buildings on climate change, improve the use of national power grids and reduce the operating expenses for owners [1]–[4]. Consequently, promoting the concrete-based envelop by adding a gainful function based on the production of electricity tends to increase the value to the residential and office buildings [5]–[10]. The installation of photovoltaics on several vertical façades of buildings could also partially answer the seasonal mismatch between the production and the consumption of electricity. Indeed, different vertical exposures would enlarge the production of photovoltaic electricity over longer periods of time (earlier/later during day, in Spring and Autumn), than in case of photovoltaics only installed on the roof, facing the sun direction in Summer (to maximize the electricity production, but over a restrained range of time).

Our researches have been undertaken by using concrete as a direct support of the solar cells/films by the way of an aesthetical cladding system. Consequently, we have mainly focused on the use of ultra-high performance fibre-reinforced concrete (UHPFRC), which is known to present a smooth and durable surface due to its mix-design and specific interactions with the formworks and demoulding agents [11]–[13]. A patented process (already described in a previous paper [14]) aims to manufacture a multi-layer's composite material based on UHPFRC, specific structural adhesives and solar cells (or films).

In this paper, we will overview the several steps of development done during the last two years and/or always in progress. After summarizing the different tests of durability done on the interface between concrete, glue and photovoltaic devices, we will present the results from the first tests of production of integrated façade panels, undertaken in a pre-cast plant

WIT Transactions on Ecology and the Environment, Vol 222,
www.witpress.com, ISSN 1743-3541 (on-line)
doi:10.2495/EQ180111

in 2017. Finally, we will detail the next step under progress, with the building of a first outdoor demonstrator near Lyon (France), in order to give proofs of durability under weathering, and to monitor their efficiency over several years.

2 DESCRIPTION OF THE PROCESS OF INTEGRATION

The process of integration of the PV cells on concrete and the composition of the raw materials (concrete, glue, PV cells) were already detailed in a previous paper [14]. They can be quickly summarized in successive steps:

- A layer of glue or an adhesive tape is applied on the backside of the PV cell/film, which is deposited at the bottom of the casting mould.
- The fresh mix-design of concrete is poured into this mould in order to coat the layer of glue/adhesive.
- After about 18 hours, the concrete has hardened and all the multilayers system can be removed from the mould.

The Fig. 1 shows some pictures of the first prototype (curved shape of 3m length) done using ultra-high performances fibres concrete, an adhesive tape composed of polyisoprene-polyisobutylene and a flexible PV film (composed of a-silicon).

3 CHARACTERISATION OF THE INTERFACE AND DURABILITY

The morphology and the composition of polished cross-sections of the multilayer composing the panels were characterized by scanning electron microscope (SEM FEG Quanta 400 from FEI Company, accelerating voltage of 15keV and current intensity of 1nA) and Energy-dispersive X-ray spectroscopy (EDS, XFlash®-100 SDD detector from Bruker). As detailed by Fig. 2, no particular air void was detected at the interface between concrete, epoxy-based glue and the backside of the PV cell. Moreover, the hydrated cement paste containing limestone-based filler and sand is characteristic of the mix-design of the concrete. As detected by EDS, the epoxy-based glue was composed of fine filler made of magnesium sulphate and barium sulphate. According to the multiple analyses done, the thickness of the layer of glue was estimated to 600μm +/-300μm.

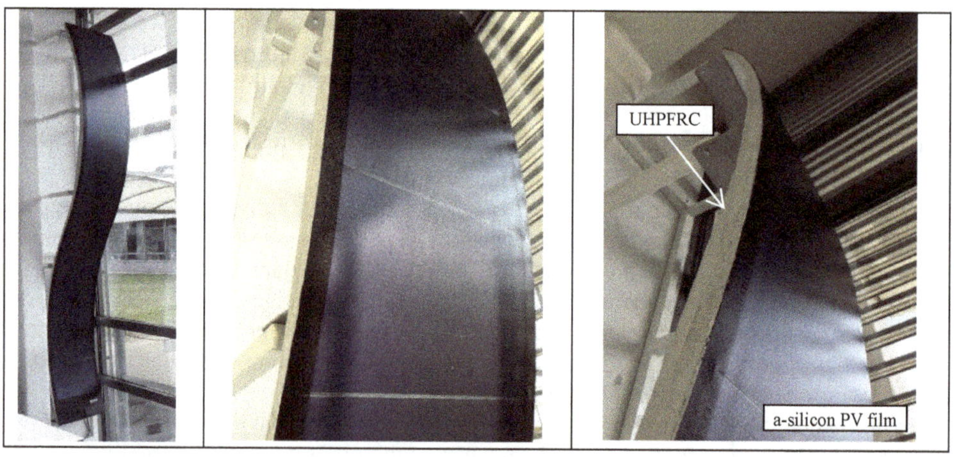

Figure 1: First prototype (curved panel) based on UHPFRC integrating a flexible 3m-long photovoltaic film.

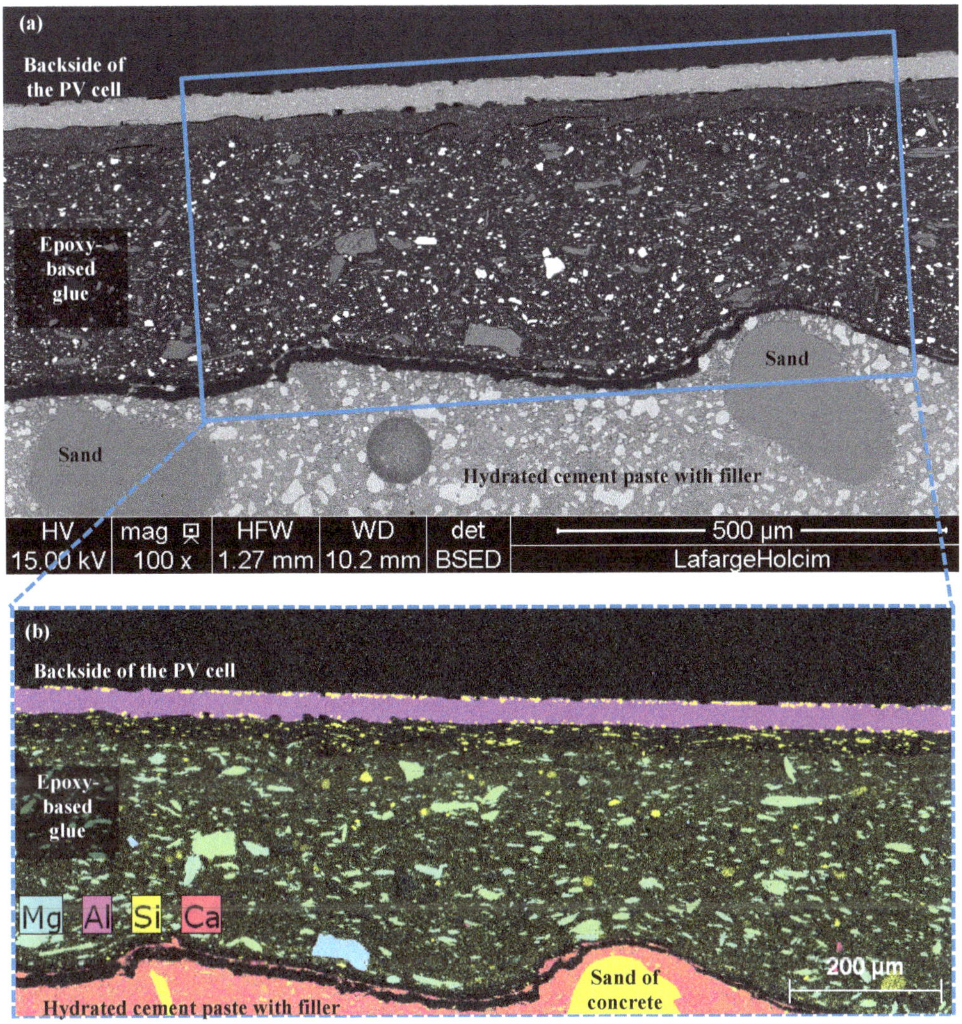

Figure 2: (a) Polished cross-section of the interface between concrete, epoxy-based glue and the backside of a PV cell: observed by SEM; (b) Analysed by EDS.

The adhesion and the durability of the interface were studied using the pulling out method (using a dynamometer) and several accelerated aging methodologies [14]:

- Continuous storage in a condensation chamber (60°C/100% relative humidity),
- Storage in a UV-rays chamber (4h at 60°C with UV, followed by 4h at 50°C without UV),
- Storage in a freeze-thaw chamber (cycling between +9°C and -18°C).

According to the tests performed (Fig. 3, [14]), the best results in terms of durability were obtained using an epoxy-based glue or an adhesive tape made of polyisoprene-polyisobutylene (until 800 freeze-thaw cycles and until 9 months of UV aging/condensation tests without any debonding issue). Moreover, it was possible to establish the best range of roughness for the backside of the PV cells/films making possible the best adhesion with the layer of glue (values of adhesion until 1.18 MPa were obtained).

4 TESTS OF INTEGRATION IN PRECAST PLANT

We decided to conduct tests in precast plant to give first proofs of industrialisation of the new PV system based on UHPFRC. The main goals were to bring a high architectural quality to the façade panels without any significant modifications of the process line. More than 30 photovoltaic façade panels were then produced during two trials performed in February and October 2017.

As shown in Fig. 4, the dimensions of each façade panel were about 2.4m length by 1.2m width, incorporating about 1.5m² of PV films (covering about 50% of the area). As shown in Fig. 4(d) (in red), the electrical connectors (junction boxes) emerge directly on the backside of the concrete-based panels, without any cable visible on its front-side.

These trials also established that this new type of photovoltaic façade panels were able to resist against water during the several post-treatments done after hardening of concrete (polishing, cleaning, cutting steps), showing a nice and smooth aesthetic for a minimum of thickness of the façade panels (less than 20 mm).

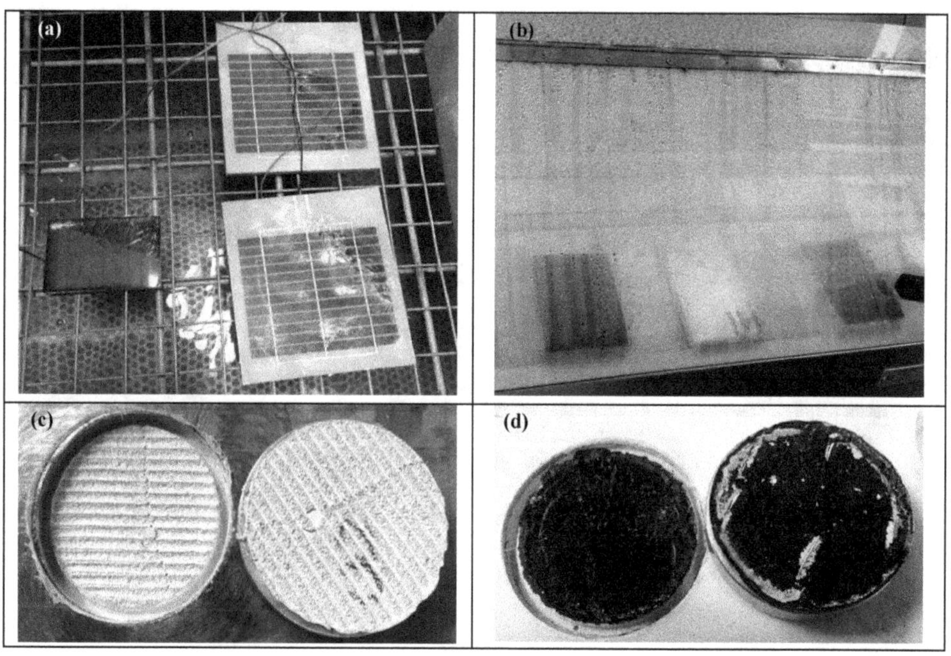

Figure 3: (a) Some durability tests and pictures of the interfaces after pulling out (adhesion) tests: samples covered by ice during freeze-thaw test; (b) Samples in the condensation chamber; (c) Interface after the pulling out test with debonding between epoxy-glue layer and concrete, (d) Interface with debonding into the adhesive tape.

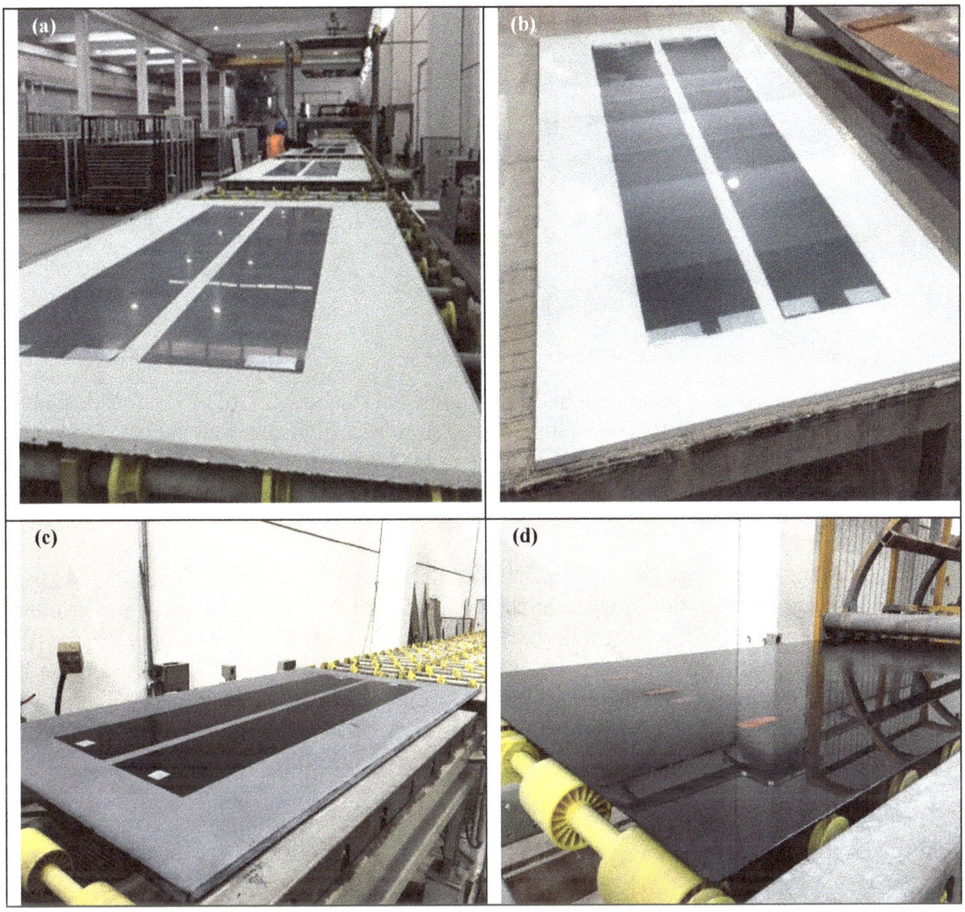

Figure 4: (a), (b), (c) Façade panels based on UHPFRC and integrating PV films; (d) Frontside backside showing the outputs for the electrical devices (in red).

5 DEMONSTRATOR AND PROSPECTIVE ACTIONS

A first demonstrator of the entire cladding system is scheduled in 2018 near Lyon (France), using some of the PV façade panels produced in precast plant. About 10 façade panels (maybe until 14), showing a total of 13m² of photovoltaic area, fully integrated on 25m² area of concrete-based façade panels, are expected to be installed on a main structure (5.0m length, 2.5m width, 2.8 m height) already built in 2017.

As shown in Fig. 5, the PV façade panels will be cladded on at least 3 walls of the structure (facing respectively to East, South and West). Moreover, no cable is expected to be visible on the front-side, keeping as more aesthetical as possible the façades of the building. The main goal of this demonstrator is to confirm that the installation of the PV façade panels based on UHPFRC is possible without any complex modifications and without over-costs. The other goals are to give proofs of durability under weathering, and to monitor their efficiency over several years and under different sun exposures.

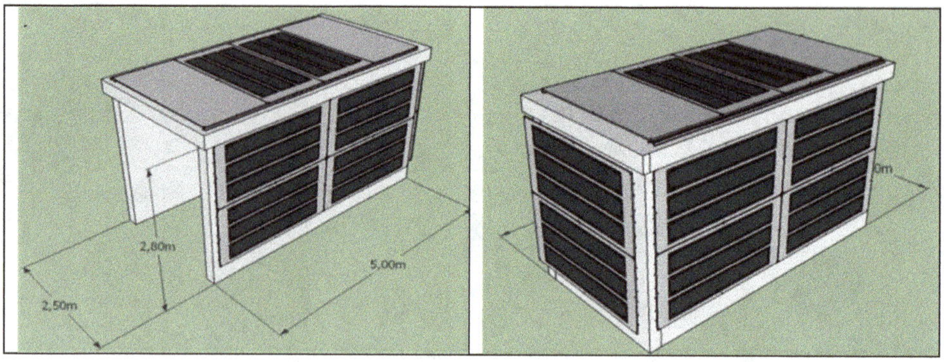

Figure 5: Demonstrator scheduled in 2018, composed of about 25 m² of PV facade panels made of UHPFRC. The main vertical façades will be exposed to East, South and West in order to monitor the aging and the efficiency of the PV films according to the sun exposure.

6 CONCLUSIONS

New solutions about Building Integrated Photovoltaic are necessary to answer the growing needs for a decentralized generation of renewable energies. Complementary to the PV installed on the roof, the development of aesthetical photovoltaic façade could be an opportunity to: (i) provide more energy to heat or cool residential and office buildings, (ii) enlarge the production of electricity over longer periods of time (all along the day and year), (iii) minimize the consumption of fossil fuels, and (iv) reduce the impact of urban areas on the climate changes.

In this study, we developed a cladding system based on ultra-high-performance fibre-reinforced concrete panels integrating PV films. After establishing the durability of the multilayers system using accelerating tests in laboratory, we showed the first proofs of industrialisation in 2017 by doing trials in precast plant and by producing more than 30 façade panels (covered at about 50% by the photovoltaic devices).

The next step is the building of a demonstrator composed of 25m² of façade panels, which is scheduled for 2018. This demonstrator will be helpful to give proofs of durability under weathering, but also to monitor the efficiency the PV films for different sun exposures (East-South-West).

ACKNOWLEDGMENT

The authors would like to thank Alexander Eisenhauer (LafargeHolcim) for his helpful work regarding the innovation and marketing strategies.

REFERENCES

[1] Alnaser, N.W., Building integrated renewable energy to achieve zero emission in Bahrain. *Energy and Buildings*, **93**, pp. 32–39, 2015. DOI: 10.1016/j.enbuild.2015.01.022.

[2] El Gindi, S., Abdin, A.R. & Hassan, A., Building integrated photovoltaic retrofitting in office buildings. *Energy Procedia*, **115**, pp. 239–252, 2017. DOI: 10.1016/j.egypro.2017.05.022.

[3] Sorgato, M.J., Schneider, K. & Rüther, R., Technical and economic evaluation of thin-film CdTe building integrated photovoltaics (BIPV) replacing façade and rooftop

materials in office buildings in a warm and sunny climate. *Renewable Energy*, **118**, pp. 84–98, 2018. DOI: 10.1016/j.renene.2017.10.091.

[4] Curtius, H.C., The adoption of building-integrated photovoltaics: barriers and facilitators. *Renewable Energy*, **126**, pp. 783–790, 2018. DOI: 10.1016/j.renene.2018.04.001.

[5] Ceron, I., Caamaño-Martin, E. & Neila, F.J., 'State-of-the-art' of building integrated photovoltaic products. *Renewable Energy*, **58**, pp. 127–133, 2013. DOI: 10.1016/j.renene.2013.02.013.

[6] Cronemberger, J., Almagro Corpas, M., Ceron, I., Caamaño-Martin, E. & Vega Sanchez, S., BIPV technology application: highlighting advances, tendencies and solutions through solar Decathlon Europe houses. *Energy and Buildings*, **83**, pp. 44–56, 2014. DOI: 10.1016/j.enbuild.2014.03.079.

[7] Polo Lopez, C., Frontini, F., Bonomo, P. & Scognamiglio, A., PV and façade systems for the building skin. Analysis of design effectiveness and technological features. *Proceedings of the Twenty-ninth European Photovoltaic Solar Energy Conference and Exhibition*, Amsterdam, 2014.

[8] Verberne, G. et al., BIPV products for façades and roofs: a market analysis. *Proceedings of the Twenty-ninth European Photovoltaic Solar Energy Conference and Exhibition*, Amsterdam, 2014.

[9] Shukla, A.K., Sudhakar, K. & Baredar, P., A comprehensive review on design of building integrated photovoltaic system. *Energy and Buildings*, **128**, pp. 99–110, 2016. DOI: 10.1016/j.enbuild.2016.06.077.

[10] Evola, G. & Margani, G., Renovation of apartment blocks with BIPV: energy and economic evaluation in temperate climate. *Energy and Buildings*, **130**, pp. 794–810, 2016. DOI: 10.1016/j.enbuild.2016.08.085.

[11] Gueit, E., Darque-Ceretti, E., Tintillier, P. & Horgnies, M., Surfactant-induced growth of calcium hydroxide coating at the concrete surface. *Journal of Coating Technology and Research*, **9**(3), pp. 337–346, 2012. DOI: 10.1007/s11998-011-9368-4.

[12] Horgnies, M. & Chen, J.J., Superhydrophobic concrete surfaces with integrated microtexture. *Cement and Concrete Composites*, **52**, pp. 81–90, 2014. DOI: 10.1016/j.cemconcomp.2014.05.010.

[13] Horgnies, M., Darque-Ceretti, E., Gueit, E. & Aucouturier, M., Characterisation of the variations of tint and the adhesion of pigments onto the surface of mortar. *Construction and Building Materials*, **94**, pp. 380–386, 2015. DOI: 10.1016/j.conbuildmat.2015.07.083.

[14] Horgnies, M., Legrand, F., Bonnet, E. & Dubois-Brugger, I., Study of the interface and durability of the adhesion between photovoltaic cells and concrete-based façade panels. *WIT Transactions on Engineering Sciences: Materials and Contact Characterisation* VIII, WIT Press, **116**, pp. 171–179, 2017.

PRODUCTION OF LIQUID HYDROCARBONS FROM MILLET HUSK VIA CATALYTIC HYDRODEOXYGENATION IN NIO/AL_2O_3 CATALYSTS

NASIR ANKA GARBA[1], IBRAHIM K. MUDURU[2], MOHAMMED A. SOKOTO[2] & SANI M. DANGOGGO[2]
[1]Department of Chemistry, Federal University Gusau, Nigeria
[2]Department of Pure and Applied Chemistry, Usmanu Danfodiyo University Sokoto, Nigeria

ABSTRACT
A number of viable options are being explored for the production of energy to meet growing global energy demand. The research investigated the conversion of millet husk to liquid hydrocarbons. Furfural was produced from millet husk via a single-stage process that involved simultaneous hydrolysis and dehydration processes, and was subsequently converted to straight-chain hydrocarbons through aldol condensation and hydrodeoxygenation reactions. The hydrodeoxygenation of the aldol condensation product over NiO/Al_2O_3 (493 K, 30 bar and reaction time of 1 hour) yielded liquid hydrocarbons with the selectivity of 77.5%. The results indicate that millet husk could be a good substrate for furfural production, and could be a source of fuels and chemicals for industries.
Keywords: biofuels, hydrocarbons, furfural, aldol condensation, hydrodeoxygenation, millet husk, catalysis, FTIR, GC-MS, biomass.

1 INTRODUCTION

As concern grows over the twin challenges of climate change and energy security, a number of viable options are being explored for the production of energy to meet the global demand in a sustainable fashion. One of these options is biofuel production from renewable biomass. The exploitation of biomass for energy and biochemical production has been recognized as a viable option for adding value to lignocellulosic agricultural residues such as corn stover, wheat straw, rice husk/straw, and millet husk [1]. Lignocellulosic biomass consists of three major components: cellulose (40–50%), hemicellulose (25–35%), and lignin (15–20%) [2].

Biomass is a renewable resource of carbon produced from organic matter including agricultural residues. These materials can be converted to energy carrier and/or heat as well as liquid fuels by either thermochemical, physicochemical, or biological processes (Fig. 1). According to Mehdi et al. [3], biomass conversion can yield biodiesel, alcohols (ethanol/methanol), biogas, biomass-to-liquid, and hydrogen.

However, the renewable production of biofuels and chemicals will solely rely on biomass, since, among all the renewable resources, biomass is the only source of carbon [3]. A complete transition from petroleum to biomass feedstocks seems feasible for the production of biofuels and chemicals based on the current production volume of biomass; however, biomass feedstocks, being highly functionalized molecules, are significantly different from petroleum feedstocks that are generally unfunctionalized, presenting significant technological, economic, and environmental challenges [4], [5].

One of the main concerns in the large-scale production of biofuels and chemicals is the consumption of edible biomass such as corn, sugarcane, soybeans, and vegetables as feedstocks for biofuels generation. This issue has motivated researchers around the world to develop second-generation technologies for processing non-edible biomass (lignocellulosic biomass) so that the sustainable production of a new generation of fuels and chemicals can be achieved without affecting food supplies. In this respect, lignocellulosic biomass is

WIT Transactions on Ecology and the Environment, Vol 222,
www.witpress.com, ISSN 1743-3541 (on-line)
doi:10.2495/EQ180121

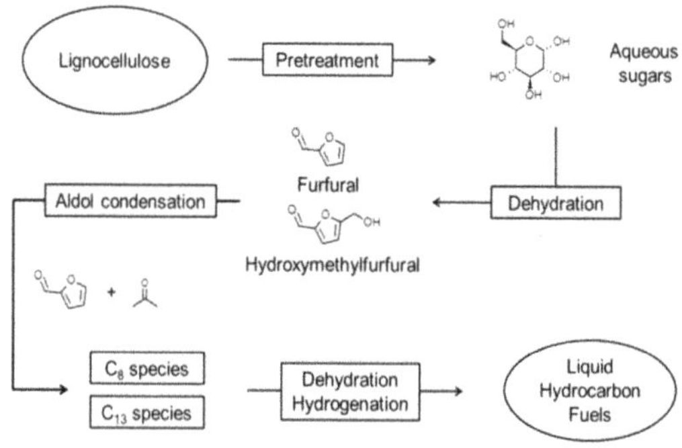

Figure 1: Biomass conversion process.

attractive since it is abundant and can be grown faster and more economically than food crops [1]. Lignocellulosic biomass consists of three major components: cellulose (40–50%), hemicellulose (25–35%), and lignin (15–20%) [2]. The aim of this research is to examine the potential of millet husk for the generation of hydrocarbons via an aqueous-phase process.

2 METHODOLOGY

Millet husk was first converted to furfural, which was further converted to hydrocarbon intermediates via aldol condensation. These were subsequently converted to liquid hydrocarbons with a selectivity of 77.5% by catalytic hydrodeoxygenation using NiO/Al_2O_3 at a 220°C reaction temperature, and 30 bar for 1 hour.

2.1 Production of furfurals

Millet husk sample was collected, cleaned, dried, and stored in a dry place in the laboratory. A 5.0 g sample of dried millet husk was prepared and 5.0 g of sodium chloride (NaCl) was added to the sample and mixed in a clean beaker. The mixture was placed in a 250 cm^3 borosilicate glass tube reactor followed by the addition of 50 cm^3 of diluted sulfuric acid (H_2SO_4). The reactor was placed upright inside a furnace and connected to a water condenser. A distillation process was then carried out based on the chosen variables of acid concentration, temperature, and reaction times. The organic portion of the distillate was then extracted with dichloromethane using a separating funnel, and the solvent was removed using a rotary evaporator at 40°C. The resultant solution was analyzed using a Fourier transform infra-red (FTIR) spectrometer [6].

2.2 Aldol condensation

The aldol condensation of furfural with acetone was performed using a flat-bottomed flask of 250 cm^3 capacity equipped with a magnetic stirrer. Then, 10 cm^3 of the distilled furfural was mixed with 5 cm^3 of acetone in the ratio of 2:1, and was subsequently transferred into the reactor, where 50% water/ethanol was added to the mixture. The reactor was heated at 85°C and 20 cm^3 of 4M NaOH was added to the mixture with vigorous stirring (500 rmp/min) for 30 min [7]. The resulting mixture was filtered and washed three times with ethanol to remove excess NaOH. Any crystalline products that had formed were dissolved in ethyl acetate. The products were analyzed using a FTIR and ultraviolet–visible spectroscopy at a wavelength of 200 nm and 400 nm, respectively [8].

2.3 Hydrodeoxygenation

Hydrodeoxygenation of the aldol product was carried out at 493 K in a 316 L stainless steel tubular reactor. Then, 30 cm^3 of the aldol adduct and 1.8 g NiO/Al_2O_3 catalyst was transferred into the reactor and the reactor was purged with nitrogen gas. The reactor was pressurized with hydrogen at up to 30 bars and heated to 493 K for 1 hour. The liquid product obtained at the end of the reaction was then filtered and analyzed by gas chromatography and mass spectrometry (GC–MS) [9].

3 RESULTS AND DISCUSSION

A Cary 630 FTIR machine was used to identify the furfural produced. The results are presented in Tables 1–3.

Table 1: FTIR spectral analysis of the produced furfural.

Class of hydrocarbons	**Functional group**	**Wavenumber (cm^{-1})**
Aldehyde	C=O	1720–1740
	C-H	2700–2900
Aromatic	C-H	3000–3100
	C=C	1450–1600
Ether	C-O	1000–1300

Table 2: FTIR spectral analysis of the aldol adducts.

Class of hydrocarbons	**Functional group**	**Wavenumber (cm^{-1})**
Ketone	C=O	1700–1725
Alkenes	C-H	3080–3140
	C=C	1630–1670
Aromatic	C-H	3000–3100
	C=C	1450–1600
Ether	C-O	1000–1300

Table 3: Compositions of hydrodeoxygenated products.

Name of hydrocarbon	**Molecular formula**	**% peak area**
Decane	$C_{10}H_{22}$	13.59
3-methyltridecane	$C_{10}H_{22}$	12.01
Tetradecane	$C_{14}H_{30}$	5.36
Pentadecane	$C_{14}H_{30}$	6.13
Hexadecane	$C_{15}H_{32}$	17.48
Hexadecane	$C_{16}H_{34}$	12.52
Eicosane	$C_{20}H_{42}$	10.41
Docosane	$C_{22}H_{46}$	13.59

As shown in Fig. 2, a strong absorption was shown by the IR spectrum at 1669 cm^{-1}, which corresponds to the absorption of conjugated carbonyl (C=O). The C=O absorption is lower than the usual absorption of aldehyde due to internal hydrogen bonding, which occurs in conjugated unsaturated aldehyde, and conjugation lowers the vibrational frequency of carbonyl compounds. The presence of the aldehyde group was proven with the existence of two weak absorptions gained at 2848 cm^{-1} and 2812 cm^{-1}, as observed in the standard furfural that indicated moderate intense stretching of aldehydic C-H, which is attributable to Fermi resonance between the fundamental aldehydic C-H stretching and the first overtone of the aldehydic C-H bending vibration [11]. Moreover, strong peaks from 1570 cm^{-1} to 1466 cm^{-1} indicated inactive stretching of C=C from the aromatic ring, and strong peaks at 1020 cm^{-1} indicated the C-O stretching vibration. These IR spectrum changes were also compared with that of the standard furfural spectrum, and that of Alonso et al. [12] and Ong and Sashikala [6].

The IR spectrum of the furfural-acetone dimer obtained is shown in Fig. 3. The frequency of the main absorption bands illustrated that the (C-H) vibrational frequency appears at 3142 and 3121 cm^{-1}, which correspond to the sp^2 vibration in the furan ring. The stretching absorbance (C=O) observed at 1603 cm^{-1} infers the carbonyl group, and a sharp stretching absorbance band at 1013 cm^{-1} probably indicates C-O stretching for cyclic C-O-C linkage of furfural. The peaks at 1553 cm^{-1} and an overtone at 2116 cm^{-1} may be as a result of C=C stretching vibration in the furan unit. Similarly, the lower broad absorption at 3401 cm^{-1} could be due to the presence of O-H vibration, which could be attributed to the absorption of the solvent that remained in the product. The spectrum observed was similar to those reported by Gharbi et al. [13] and Gandini [14].

The results presented in Table 3 show the distribution of the alkanes obtained from the hydrodeoxygenation of aldol adducts. The products were mainly liquid hydrocarbons with a carbon range of C_{10}–C_{22}. The total relative peak area of the products was 77.50%, with hexadecane and decane being the major products, each of which accounts for 17.48% and 13.59%, respectively. Based on GC–MS data, the liquid phase also contains significant amounts of two other alkanes larger than C_{16} – eicosane and docosane – which could be formed due to a decomposition and polymerization reaction. In the liquid phase, other than alkanes, there were some other organic oxygenates identified by the GC–MS, including ethanol, 2-oxepanone, and acetic acid, which are most likely byproducts. The results obtained

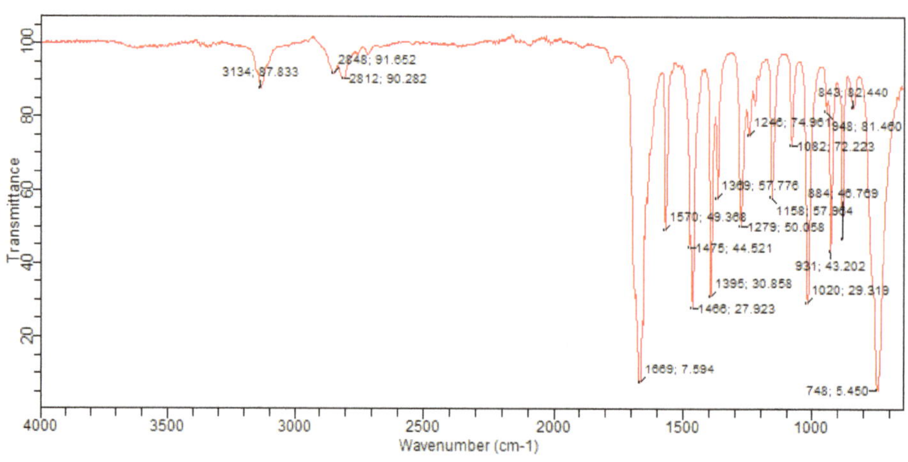

Figure 2: FTIR spectrum of the furfural produced from millet husk.

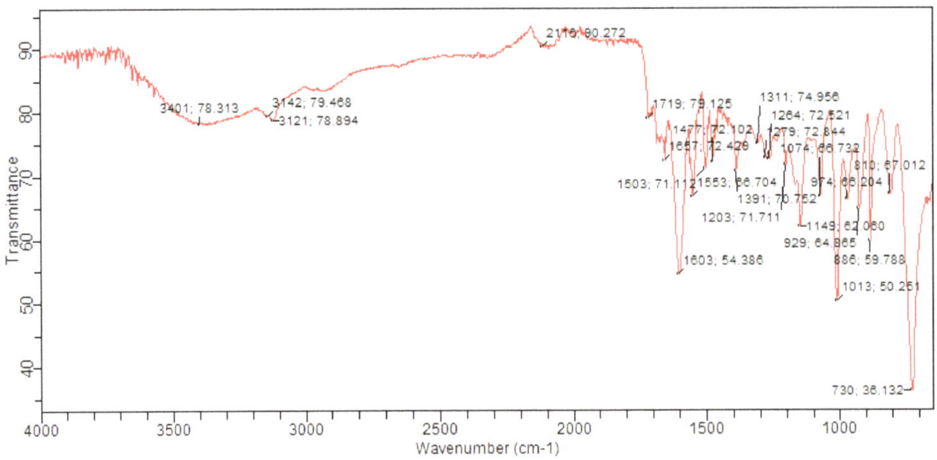

Figure 3: FTIR spectrum of the aldol adduct.

by Rong et al. [15] showed an overall experimental yield of 76% liquid alkanes (jet fuel range), which corresponds to a weight-percent yield of 0.46 kg of alkanes per kg of xylose (monomer and oligomers) in the hemicelluloses extract. Other results obtained by Faba et al. [9] also reported 21.5% of tridecane from the hydrodeoxygenation of furfural-acetone condensation adducts over Pt/Al_2O_3 catalysts. The hydrodeoxygenation step could also be improved with improvements in the catalysts and reactor design. The straight alkanes produced could be further upgraded via the hydroisomerization process to form branched alkanes.

4 CONCLUSION

The study demonstrated the conversion of millet husk to liquid hydrocarbons via an aqueous phase process. Millet husk was converted to furfural and the produced furfural was further converted to hydrocarbon intermediate via aldol condensation, which was subsequently converted to liquid hydrocarbons with selectivity of 77.5% by catalytic hydrodeoxygenation using NiO/Al_2O_3 at a 220°C reaction temperature, 30 bar for 1 hour. Thus, millet husk could be a good substrate for furfural production and could be a source of biofuel and biochemical for industries. The transformation potential of millet husk provides a new and useful process for the development of biofuels from waste lignocelluloses biomass.

REFERENCES

[1] Roman-Leshkov, Y. & Dumesic, J.A., Solvent effects on fructose dehydration to 5-hydroxymethylfurfural in biphasic systems saturated with inorganic salts. *Top Catal*, **52**, pp. 297–303, 2009.

[2] Gurbuz, E.I., Alonso, D.M., Bond, J.Q. & Dumesic, J.A., Reactive extraction of levulinate esters and conversion to γ-valerolactone for production of liquid fuels. *ChemSusChem*, **4**(3), pp. 357–361, 2011.

[3] Mehdi, H., Fabos, V., Tuba, R., Bodor, A. & Mika, L.T., Integration of homogeneous and heterogeneous catalytic processes for a multi-step conversion of biomass: from sucrose to levulinic acid, γ-valerolactone, 1,4 pentanediol, 2-methyltetrahydrofuran and alkanes. *Green Chem*, **10**, p. 238, 2008.

[4] Rao, R.S., Baker, R.T. & Vannice, M.A., Furfural hydrogenation over carbon-supported copper. *Catal Lett*, **60**(1–2), pp. 51–57, 1999.

[5] Kunkes, E.L., Simonetti, D.A., West, R.M., Serrano-Ruiz, J.C., Gartner, C.A. & Dumesic, J.A., Catalytic conversion of monofunctional hydrocarbons and targeted liquid-fuel classes. *Science*, **322**, pp. 417–421, 2008.

[6] Ong, H.K. & Sashikala, M., Identification of furfural synthesized from Pentosan in rice husk. *J Trop Agric & Fd Sc*, **35**(2), pp. 305–312, 2007.

[7] Fakhfakh, N., Cognet, P., Cabassud, M.Y. & Rioios, M.D., Stoichio-kinetic modeling and optimization of chemical synthesis: application to the aldolic condensation of furfural on acetone. *Chem Eng Process*, **47**(3), pp. 349–362, 2008.

[8] Martinez, A., Rodriquez, M.E., York, S.W., Preston, J.F. & Ingram, L.O., Use of UV absorbance to monitor furans in diluted acid hydrolysates of biomass. *Biotechnol Prog*, **16**, pp. 637–641, 2000.

[9] Faba, L., Diaz, E., Vega, A. & Ordonez, S., Hydrodeoxygenation of furfural-acetone condensation adducts to tridecane over platinum catalysts. *Catalysis Today*, **269**, pp. 132–139, 2015.

[10] Choudhary, V., Sandler, S.I. & Vlachos, D.G., Conversion of xylose to furfural using Lewis and Bronsted acid catalyst in aqueous media. *ACS Catal*, **2**(9), pp. 2022–2028, 2012.

[11] Pavia, D.L., Lampman, G.M. & Kriz, G.S., *Introduction to Spectroscopy*, 3rd ed., Thomson Learning Inc., 2001.

[12] Alonso, D.M., Wettstein, S.G., Bond, J.Q., Root, T.W. & Dumesic, J.A., Production of biofuels from cellulose and corn stover using alkylphenol solvents. *ChemSusChem*, **4**, pp. 1078–1081, 2011.

[13] Gharbi, S., Afli, A. & Gandini, A., Polyamides incorporating furan moieties. Synthesis, characterization and properties of a homologous series. *Polymer International*, **50**, p. 509, 2001.

[14] Gandini, A., Furans as offspring of sugars and polysaccharides and progenitors of a family of remarkable polymers: a review of recent progress. *Polymer Chemistry*, **1**, p. 245, 2010.

[15] Rong, X., Ayyagari, V., Subrahmanyam, H.O., Wei, Q., Peter, V.W., Hemant P. & Huber, G.W., Production of jet and diesel fuel range alkanes from waste hemicellulose-derived aqueous solutions. *J RSC*, **140**, pp. 77–84, 2010.

CHALLENGES AND OPPORTUNITIES FOR ENERGY RECOVERY FROM MUNICIPAL SOLID WASTE IN THE RUSSIAN FEDERATION

STEPAN POLYGALOV[1], GALINA ILINYKH[1], NATALIA SLIUSAR[1], VLADIMIR KOROTAEV[1] & IRINA RUKAVISHNIKOVA[2]
[1]Environmental Protection Department, Perm National Research Polytechnic University, Russia
[2]Environmental Economics Department, Ural Federal University, Russia

ABSTRACT

Energy recovery from municipal solid waste (MSW) is currently a key factor in waste management in Russia, as can be seen from the corresponding changes in the state regulatory framework. The paper presents a comparative assessment of different waste flow characteristics; it focuses on pre-treatment options and the method of subsequent thermal treatment. It reflects the basic method of untreated MSW incineration and other advanced technologies which have different levels of waste preparation. Step-by-step research has been carried out in a large Russian city in order to determine different technological approaches to recovering energy from MSW. The study included: 1) field studies of the composition and characterization of MSW; 2) laboratory studies of MSW thermal properties, i.e. moisture, ash content and calorific value; 3) several options for waste treatment schemes with equipment of different technical specifications; 4) analysis of material flow, energy flow and key substance flow according to these considered schemes. The research revealed that for all seasons a significant proportion of MSW (almost 40%) is represented by fines (below 50 mm). On average, the energy potential of MSW (the total content of waste paper, polymers, textiles, wood, multilayer packaging, leather, rubber, footwear) amounts to about 40%. It was found that MSW calorific value depends on fraction size, since screening (over 50 mm) is characterized by high moisture and low content of valuable components, compared to MSW greater than 50 mm in size. Pre-processing and treatment of MSW allows solid recovered fuel (SRF) to be obtained with a calorific value similar to brown coal. The energy balance of the process allows the conclusion that 33.8% of initial MSW energy potential is concentrated in residues from the sorting process, since it has a relatively low calorific value of 4.8 MJ/kg, but a high mass fraction of 47.9%. In the process of MSW treatment and solid fuel production, most of the chlorine and mercury (80% and 55% respectively) is removed by the rest of the sorting process. The data obtained in the study can be used in selecting the optimal option for energy recovery from MSW.

Keywords: municipal solid waste, energy recovery, solid recovered fuel, calorific value.

1 INTRODUCTION

MSW includes components with combustible properties which can be used for the industrial production of energy. In many countries of Northern and Western Europe, recovery of energy from MSW ranges from 20% to 50% [1]: 53% in Norway; 51% in Denmark; 48% in Switzerland; 55% in Finland, and 50% in Sweden. France, Germany and the Netherlands account for more than half of European incineration capacity; the total share of burning MSW in Italy, Sweden and the UK is 74% [2]; Japan reportedly burns 78% [1]; and Taiwan and Singapore a similar amount. There are different methods and technologies for energy recovery from MSW, the most common being direct waste combustion without preliminary preparation at waste-to-energy plants (WTE). The alternative solution is MSW treatment on the waste sorting line, where secondary raw materials are separated and solid fuel with high calorific value is produced. This can be used later in various energy-consuming processes, in particular, in cement kilns.

European countries mainly produce solid fuels from waste by mechanical or mechanical-biological treatment; in this case different processes for waste treatment are used [3]. Papers

WIT Transactions on Ecology and the Environment, Vol 222, © 2019 WIT Press
www.witpress.com, ISSN 1743-3541 (on-line)
doi:10.2495/EQ180131

[4]–[7] analyze incoming and outgoing flows in the production of refused derived fuel (RDF) and determine their material, energy and elemental balances.

The authors [8] have obtained high quality RDF by sorting materials from different waste flows in Singapore. They research the thermal properties and elemental composition of various waste components and discuss the suitability of solid fuel produced from MSW. In addition, they identify components with a high heavy metal content. The proximate and ultimate analysis and the calorific value of MSW components determine the quality of a combustible component, e.g. paper and food waste have a high oxygen content; there is a low content of ash and fixed oxygen in PE, PP and PS; PET has a high carbon content, but a low hydrogen value. The chlorine content in PVC is about 55%. The research has proved that the components PE, PP and PS have the highest combustion values [9].

Several studies have been carried out [10], [11] describing solid fuel values depending on MSW composition and its thermal and chemical values.

There are relatively few examples of MSW thermal treatment plants in the RF. At the JSC "Autopark No. 1 Spetstrans", St. Petersburg, SRF with the lowest calorific value of 15.9–20.3 MJ / kg is produced and it is used at a cement plant [12]. In Moscow, there are three waste incineration plants (WTE) which generate electricity; their power is relatively small (3.6–12 MW). The WTE plants in Vladivostok, Murmansk, Pyatigorsk, and Cherepovets produce thermal energy [13].

The literature review shows that in Russia there is almost no scientific research on MSW thermal properties which can be used to estimate the energy values of various waste flows obtained in the production of SRF from MSW. The authors of the article [14] investigate the methods of solid municipal waste separation in the production of solid fuel for the cement industry. They identify MSW components to be removed during the production of fuel. The article [15] looks into possible ways of using SRF in industrial furnaces for various manufactures, which are mainly run on traditional types of SRF. The assessment of MSW energy potential on stages of waste management and the assessment of thermal properties of MSW components are carried out in the city of Perm [16].

Recently, public authorities, industry and business in Russia have shown great interest in energy recovery from MSW treatment. The use of waste energy potential can solve the urgent problem of the transition from waste disposal at landfills to waste energy recovery. In this regard, the Russian Federation State Program "Energy Efficiency and Energy Development" was approved by RF Government Decree # 321 of April 15, 2014 (issued on March 30, 2013); the program includes the subprogram "Development of renewable energy sources". One of these renewable energy sources is MSW as fuel. At the end of February 2017, the Government of the Russian Federation passed a decree regulating the competitive selection and implementation of energy recovery from MSW projects. Similar to renewable energy projects, these are supported by the government through contracts in the electricity and power market. Energy production from MSW is one of the priorities in waste management; this can be seen in changes in the regulatory and legal framework of Russia, in particular in the latest development of the Federal Law on waste management. Several Federal State Standards for SRF and RDF production have been developed, specifying that the main parameters reflecting the qualitative characteristics of fuel are a lower calorific value and a content of chlorine and mercury. However, the existing waste management system in the Russian Federation presents obstacles in meeting the standards without MSW pre-treatment.

Methods and technologies for increasing MSW calorific value could include: extraction of poorly combustible components (food waste, glass, fines, etc.) using mechanical and sensor-based sorting methods; addition of high calorific value components (for example, other waste with higher calorific value); and decreasing moisture by drying.

The disadvantage of the first approach is a significant reduction in waste mass that is subject to energy recycling, and the emergence of undesirable flows that require other methods of utilization. The second method is theoretically feasible, but in practice it might be difficult to choose economically expedient high-calorific flow of materials that can be disposed of together with poorly combustible waste. Drying of waste is advisable only in cases where excess heat is applicable for other processes.

In choosing the most effective method of MSW treatment, it is recommended to take into account the energy (heat and technical) properties of waste and to justify the efficiency of proposed technologies from both points of view: required investment and the ecological effect, adjusted for the quantity and quality of disposed waste and for equipment efficiency.

2 METHODS AND MATERIALS

Evaluation of the proposed technologies for energy recovery for MSW has been carried out step by step with a specific reference to a large city in Russia and includes the following stages:

- field studies of MSW components and fractional composition;
- laboratory studies of MSW thermal properties: moisture, ash content and calorific value;
- several waste treatment schemes have been proposed with allowance for operational efficiency of equipment;
- analysis of material, energy and key substances flows according to the proposed scheme.

2.1 Investigations of MSW component and fractional composition

The methodology assessing MSW energy potential has been developed [17], [18], which includes detailed field studies of waste fractional and component composition, as well as laboratory studies of moisture and ash content of individual components. The accuracy and reliability of the results has been ensured by experimental studies of MSW components and fractional composition conducted in each season (spring, summer, autumn, winter) over 7 days, with a daily sampling and analysis of 5 samples. A sample weight amounted to 100 ± 20 kg, depending on the objectives of the studies.

To determine MSW fractional composition, several sizes of sieves were used: 250, 100, 50, 15 mm. A total list of identified components included 41 names. The studies of MSW composition included a category "water" found in bottles, jars and containers, as its evaporation consumes heat and thereby reduces the total calorific value of MSW.

2.2 Laboratory studies of MSW thermal properties

Moisture and ash content sampling were carried out in accordance with GOST 33626-2015 "Solid fuel from household waste. Sampling methods" (EN 15442: 2011). Laboratory studies of moisture content in MSW components were carried out in accordance with the requirements of GOST 33512.3-2015 "Solid fuel from household waste. Moisture specification by drying. Part 3. Analytical moisture" (EN 15414-3: 2011). Ash content of MSW components was calculated according to GOST 33511-2015 "Solid fuel from household waste. Specification of ash content" (EN 15403: 2011).

To calculate MSW net calorific value, the following formula is used [19]:

$$Q_i^r = \sum[K_n^r(1 - W_{tn}^r/100)(1 - A_n^d/100) \times Q_{in}^{daf}] - 0.02442 \times \sum(K_n^r \times W_{tn}^r), \quad (1)$$

where:

Q_i^r = calorific value of MSW, MJ/kg;
K_n^r = mass fraction of a component, %;
W_{tn}^r = moisture content of a component, %;
A_n^d = ash content of a dry mass, %;
Q_{in}^{daf} = net calorific value of a dry ashless mass, MJ/kg.

2.3 MSW treatment options

The paper focuses on the most common MSW treatment options which include MSW preparation and thermal treatment to generate heat and electric energy or to produce solid fuel from waste. Energy recovery from waste is carried out at incineration plants either by direct incineration of waste without pre-treatment, or by burning rests from sorting plants. Solid fuel from waste with a high energy value is also obtained on waste sorting lines by extracting secondary raw materials for their subsequent use in energy-consuming technological processes, e.g. in cement kilns. It is theoretically possible to increase the calorific value of separated thermally recoverable fractions by adding sub-standard secondary raw materials and combustible wastes, such as oil production and oil refinery waste or wood waste. The study does not focus on this method since it is rarely feasible.

The main objective of the current study is to conduct a comparative evaluation of waste flows focusing on the methods of pre-treatment and further thermal treatment. In this regard, the considered options include a basic method of untreated MSW incineration and some advanced methods with different levels of waste pre-treatment.

2.4 Analysis of material, energy and elemental flows

Material, energy and elemental balances of different options for MSW treatment and disposal are based on material flow analysis methodology [20]. The results are presented in STAN 2.5.1302 and Sankey diagrams. The chlorine and mercury elemental balance was calculated based on dry weight concentrations of these substances in individual components [7], [21], [22] and the composition of each flow.

3 RESULTS AND DISCUSSION

3.1 Component and fractional composition of MSW

The current research of MSW composition throughout the seasons presents an average annual calculation of MSW fractional composition and its component composition by 13 categories of components, as well as moisture value (W), ash content of individual components to dry matter (Ad) and their calorific value (CV) to dry ashless mass (Q^{daf}) (Table 1). A significant part of MSW (almost 40%) is represented by fine fraction (< 50 mm) with high moisture value and low content of valuable components. A major part of MSW composition comprises food waste, screenings, polymers and waste paper. On the basis of MSW component composition calculation, it can be concluded that MSW energy potential (the total content of waste paper, polymers, textiles, wood, multilayer packaging, leather, rubber, footwear) averages at about 40%.

Table 1: Composition and thermal properties of individual components.

Name	Composition					W, %	A^d, %	Q^{daf}, MJ/kg
	> 250 mm	100–250 mm	50–100 mm	< 50 mm	Total			
Organic waste	0.34	7.69	5.18	10.42	23.63	72.1–81.1	15.0–22,8	18.3
Waste paper	1.58	4.53	2.26	2.94	11.31	30.0–41.0	12.9–17.1	16.9
Polymers	0.66	10.17	4.10	1.75	16.69	24.2–27.0	3.8–9.9	34.5
Glass	0.00	5.83	1.42	0,39	7.64	3.5	100.0	0.0
Metal	0.12	0.89	0.36	0.03	1.40	3.5	100.0	0.0
Textile	0.86	1.52	0.51	0.04	2.92	11.5–37.9	2.7–9.3	22.6
Wood	0.47	0.46	0.13	0.02	1.08	10.5–24.3	3.0–8.5	18.9
Complex products	0.32	0.94	0.40	0.00	1.67	9.7–13.6	28.8–36.6	24.3
Hazardous substances	0.00	0.28	0.14	0.03	0.45	3.5	50.0	20.1
Inerts	1.36	1.20	0.75	1.64	4.95	4.1–8.2	100.0	0.0
Other	0.33	4.28	1.66	2.85	9.11	31.0–40.2	20.5–34.6	22.3
Fines	0.00	0.00	0.00	18.69	18.69	45.4–65.3	50.1–63.9	20.1
Additional (water)	0.14	0.26	0.05	0.00	0.45	100	0.0	0.0
TOTAL	6.18	38.04	16.97	38.81	100			

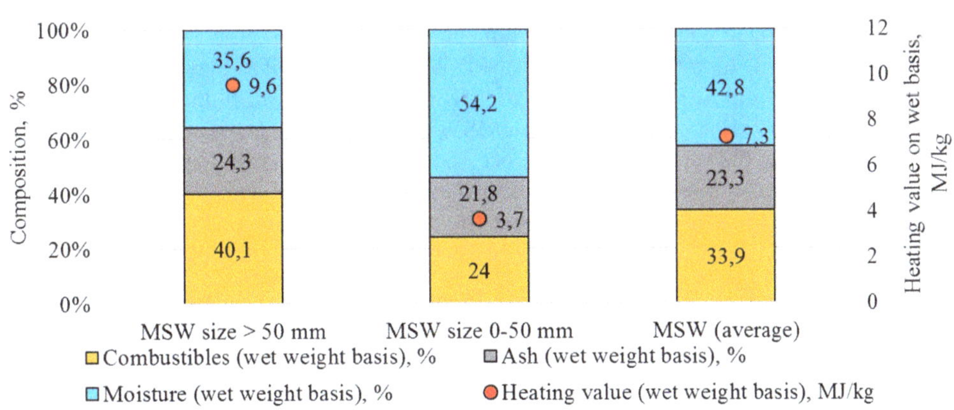

Figure 1: Thermal properties of MSW by fractions.

3.2 Thermal properties of MSW

Studies of the humidity and ash content of MSW different fractions and data on the calorific value of individual components allows for a separate calculation of calorific value for MSW fractions > 50 mm, fractions < 50 mm and for MSW as a whole. The thermal properties of MSW fractions of different sizes are presented in Fig. 1.

3.3 MSW thermal treatment options

The choice of a thermal treatment scenario has been based on the following considerations:

- The current system of MSW treatment. In Russia, the system of separate waste collection is poorly developed, though attempts have been made to implement it in some cities. However, the process is complex and lengthy, and requires major changes in the waste management system. So, all options are built on the premise that a separate waste collection system does not exist or is ineffective, as if the proposed technologies are to be introduced in the shortest possible time.
- The MSW energy recovery system, supported by regulatory legal acts. Currently, two options are considered at the legislative level: direct incineration and SRF and RDF production and consumption.
- The existing market in MSW treatment equipment and technologies.
- Studies of MSW composition and thermal properties, e.g. dependency of moisture and ash content of individual components and MSW energy value on the size of a fraction. Due to the lack of separate waste collection, MSW contains a significant proportion of organic components – food residues and plant waste, whose high moisture value affects the overall waste moisture value and, correspondingly, low heat value which is often insufficient to support self-burning of waste or required efficiency in a process [23].

Taking into account the highlighted aspects, several possible scenarios for waste pre-treatment and thermal treatment have been considered:

- Scenario 0. Direct incineration of untreated waste (S.0). It is carried out at waste-to-energy plants with generation of heat and electric energy. All waste coming to plants is incinerated without any pre-treatment.
- Scenario 1. Incineration of rests of MSW sorting (S.1). The aim is to separate fine fraction from MSW and dispose of it. The rests are sent for further incineration.
- Scenario 2. Energy recovery from MSW (S.2). Compared to the previous scenario, rests from the sorting plant are processed (sensor-based sorting) with the release of burning components. The separated thermally recoverable fractions already correspond to solid fuel from waste (by heat value, chlorine and mercury content), but do not meet the requirements of consumers, e.g. cement producers, in moisture value and particle size.
- Scenario 3. Solid fuel from waste (S.3). This option complements the previous one with necessary pre-treatment technologies to meet the requirements of customers, e.g. cement plants [24]. Shredding and drying are added to the process.

MSW pre-treatment and energy recovery technology is successive for all 4 scenarios. It is presented as a general scheme of material flows (Fig. 2), where each follow-up option is added to the previous one.

The MSW treatment and energy recovery process consists of the following operations:

- Screening (a cell size is 50 mm) is designed to sort out fines. Screening is an important stage in MSW preparation and treatment; the moisture value of fine screenings < 50 mm is higher than of those > 50 mm (Fig. 1). Fine fraction separation increases MSW calorific value.

- Separation of secondary raw materials: cardboard, office paper and newspapers, polyethylene (PE) film, PET and PE bottles, glass, ferrous and non-ferrous metals.
- The efficiency of the MSW sorting process is calculated based on operated waste sorting facilities.
- Selection of thermally recoverable fractions, aimed at the release of combustible components (waste paper and polymers), on condition that sorting efficiency is 95%. PVC components are removed from fuel to meet chlorine requirements in SRF.
- SRF is treated by shredding and drying separated thermally recoverable fractions. At this stage fuel is treated to meet consumers' quality requirements.

3.4 Evaluation of material, energy and elemental flows

The total material balance of the technological process of MSW preparation and energy recovery is calculated for a 100,000-tonne process. Fig. 2 presents the material balance of the process and shows that the distribution of components for the considered options affects the calorific value of a specific MSW flow:

- in Scenario 0, all 100% of MSW with a lower heating value of 7.3 MJ/kg can be sent to WTE plants;
- in Scenario 1, the rests of sorting process (62% of MSW initial mass) with a lower heating value of 7.7 MJ/kg can be sent to WTE plants
- in Scenario 2, thermally recoverable fractions (14.2% of MSW initial mass) with a heating value of 17.8 MJ/kg are extracted from the rests of the sorting process;
- in Scenario 3, solid fuel after drying has a heating value of 22.0 MJ/kg and accounts for 11.7% of MSW initial mass.

In scenarios S.1–S.3, incoming MSW flow generates about 11% (by mass) of secondary raw materials for further recycling and 27% of fines are sent to waste disposal sites. Scenarios S.2 and S.3 produce 48% of sorting residues, which are also sent to MSW landfills. The initial moisture value of separated thermally recoverable fractions (S.2) accounts for 25.6%; after drying and crushing, it is possible to obtain 11.7% of SRF with a moisture value of 10%. Loss of moisture at drying (S.3) amounts for 2.46%.

At the secondary raw materials stage, components with energy potential are selected, e.g. corrugated cardboard, office paper and newspapers, PE film, PET and PE bottles. In different technological processes combustible components may be extracted as thermally recoverable fractions to increase the mass fraction of waste-to-solid fuel and to increase its calorific value. Fuel quality composition is assessed by moisture value, ash content and chemical elements values: carbon, hydrogen, oxygen, nitrogen and sulphur. The presence of moisture in MSW reduces the number of combustible elements, and because heat is used to evaporate moisture during combustion, the moisture content of the waste impacts its heating value. Ash content reduces calorific value and considerably complicates the combustion process and furnace operation. Calorific value is largely determined by the content of "combustible" elements in waste: carbon, hydrogen and sulphur. Combustible elements are bound with non-combustible elements – oxygen and nitrogen – which form an internal ballast in MSW components. Table 2 compares the elemental composition and main thermal properties of waste flows for all scenarios against some fossil fuels. Table 2 shows that the content of combustible elements in the working mass is 2.4 times higher in prepared fuel than in initial MSW. Moreover, the

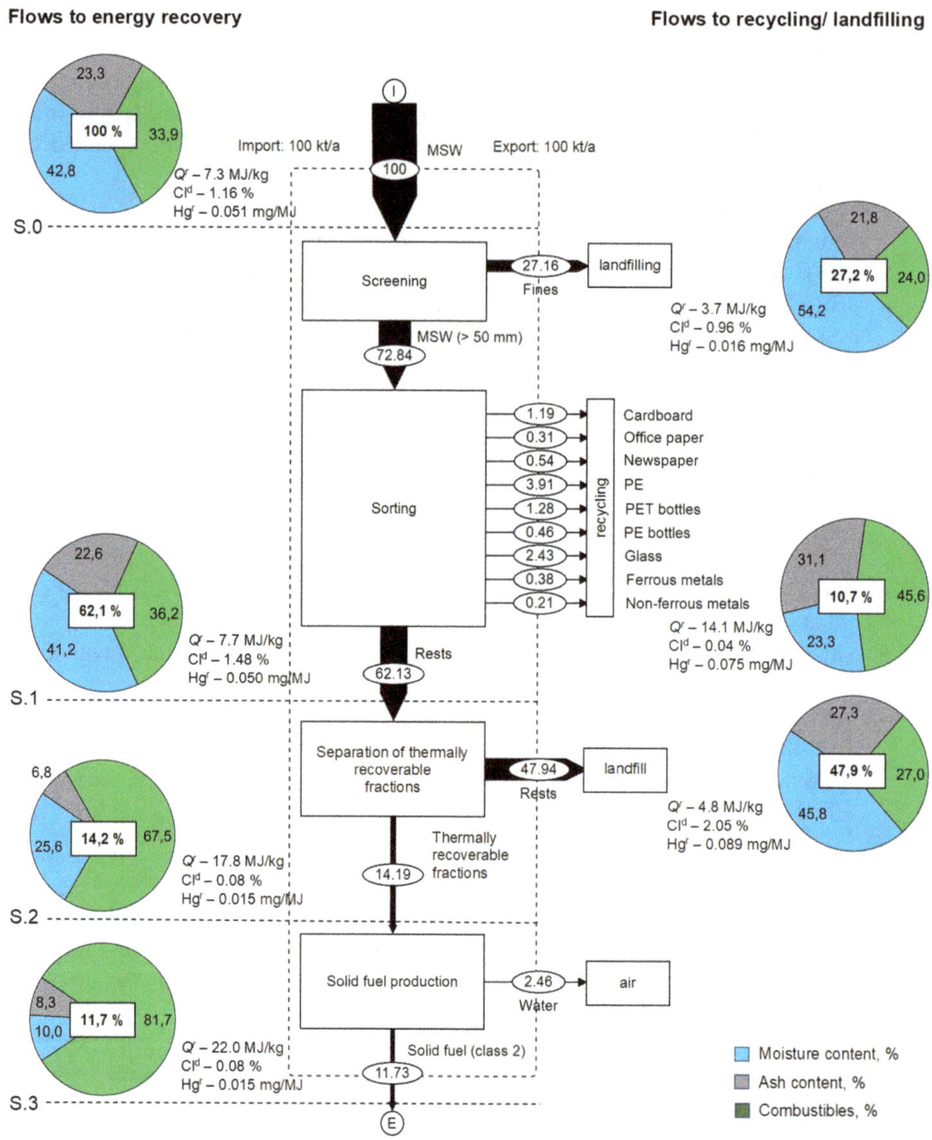

Figure 2: Material balance of MSW treatment and energy recovery.

quality of combustible fraction slightly improves as the content of calorific carbon increases and the content of ballast oxygen decreases. Thus, solid fuel from MSW is comparable to brown coal in the balance of elements.

Summarising the results, the energy balance and elemental balance for chlorine and mercury in the production of solid fuel from MSW were calculated. Fig. 3(a) presents the material balance of pre-treatment and energy recovery from MSW. The energy balance of MSW initial flow was calculated in relation to the calorific value and mass fraction of each output (Fig. 3(b)).

Sorting residues take about one-third of the energy potential of initial MSW flow, which is comparable to the energy potential of SRF. This is because sorting residues which have a relatively low calorific value of 4.8 MJ/kg surpass solid fuel more than three times in mass. Solid fuel flow has a high calorific value at a relatively small mass.

Table 2: The elemental composition and main thermal properties of waste flows for all scenarios compared to some fossil fuels.

Flow	Moisture, %	Ash content, %	Combustible mass, %	Elemental composition (on fuel mass), %					CV, MJ/kg
				C	H	O	N	S	
MSW pre-treatment and after-treatment									
MSW initial flow (S.0)	42.8	23.3	33.9	58.5	7.9	32.3	1.1	0.2	7.3
Reject material (S.1)	41.2	22.6	36.2	58.1	7.7	32.6	1.3	0.3	7.7
Thermally recoverable fractions (S.2)	25.6	6.8	67.5	62.8	8.5	28.1	0.3	0.2	17.8
SRF (2 class) (S.3)	10.0	8.3	81.7	62.8	8.5	28.1	0.3	0.2	22.0
Traditional fossil fuels for comparison [25]									
Wood	30–60	<1	40–70	49–51	5–7	38–42	0.5–2.3	–	8–10
Peat	40–55	4–8	37–56	50–60	5–7	28–33	1.5–2.5	0.2–0.3	8.1–10.5
Brown coal	0–55	7–45	0–93	64–78	4–6	15–26	0.6–1.6	0.3–6.0	7.8–19
Hard coal	5–8.5	15–20	71.5–80	89–94	2–3	1–2	0.5–1.5	2–3	23–25

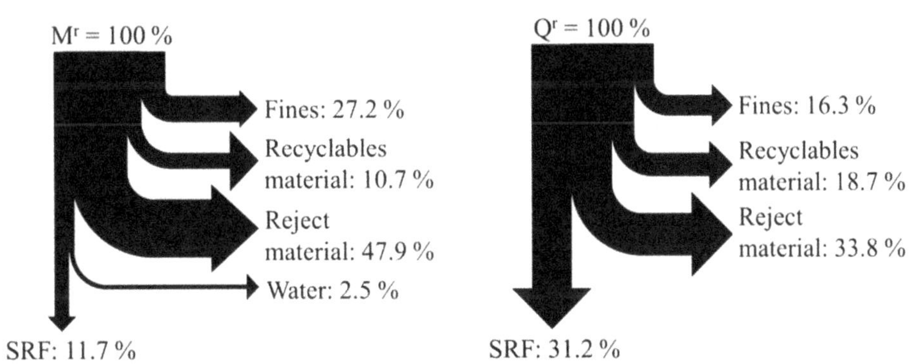

Figure 3: (a) Material; (b) and energy balances in a technological process.

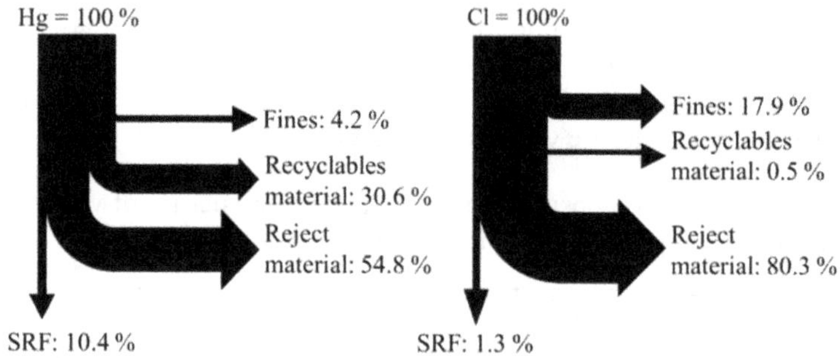

Figure 4: (a) The elemental balance for chlorine; and (b) Mercury.

The elemental balance for chlorine and mercury in original MSW flow was calculated with respect to the chlorine and mercury content of individual components, the component composition of individual flows and the mass fraction of each output (Fig. 4) [7], [21], [22].

In MSW treatment and SRF and RDF production most of the chlorine is removed to sorting residues, since chlorine is predominantly present in PVC, which is removed by sensor-based sorting. Food waste and PVC components contribute most to fines from the sorting process, where the concentration of food waste with insignificant chlorine is high, while the presence of PVC with high chlorine is low. Chlorine content in food waste can be accounted for by the presence of salt in food [7].

Mercury concentrates mainly in sorting residues and accounts for approximately 55%; the greatest contribution is made by textiles and ferrous metals [7], [22]. In the flow of secondary materials, the bulk of mercury content comes from metals (total contribution to secondary raw materials is about 78%). The content of chlorine and mercury in solid fuels from waste (S.3) meets the requirements of GOST 33516-2015 (EN 15359: 2011). Thus, selecting technologies for MSW treatment it is possible to control the flow of waste and its thermal properties, taking into account the technological characteristics of the energy recovery equipment.

4 CONCLUSIONS

Unlike many countries, in Russia the system of separate waste collection is not actually applied; research proves that there are many organic components – food and plant residues, in waste. The high humidity of these components affects the overall high humidity and relatively low calorific value accordingly. Glass as a secondary resource is not in demand, so all consumed glass is found in waste flows, again reducing overall calorific value. Cardboard, which could increase the combustible properties of waste, is in high demand in Russia as it is recycled material, so it is intensively recovered at all stages of waste treatment. Therefore, mixed MSW usually has a low calorific value which is often insufficient to maintain self-combustion or ensure the necessary efficiency of a process.

MSW has energy potential that can be controlled by selecting different treatment technologies and achieving the required calorific value of a target flow (SRF), with reference to the component composition of original MSW and its thermal properties. The latest reliable experimental data on waste energy properties highlights the potential of thermal methods of waste recovery. Laboratory studies of moisture and ash content in individual components of different MSW fractions reveal the change in MSW calorific value depending on fraction size: larger fractions have lower humidity, higher combustible mass and, correspondingly, higher calorific value.

The energy balance of a process depends on combustible components (waste paper, polymers) and their treatment: the more components are recovered as secondary raw materials, the less calorific sorting residues will be, resulting in a smaller amount of solid fuel generated. The calorific value of initial MSW flow is rather low (7.3 MJ/kg), which may present certain technological difficulties, but the combustion of MSW without preliminary treatment allows to minimize land disposal (only ash received after burning is sent to landfill). Preliminary waste sorting with separation of secondary raw materials increases the calorific value of rests from the sorting process to initial MSW flow values; however, its flow mass accounts for only 62% and it generates a flow of fine fraction that must be disposed of in landfill. When thermally recoverable fractions (S.2) are sorted from the rests, their calorific value (17.8 MJ/kg) is much higher than the calorific value of initial MSW, but this fraction averages to only 14.2% of initial MSW mass and it boosts the amount of waste sent to MSW landfill (75%). To meet the requirements for SRD and RDF, it is advisable to introduce a stage of additional preparation of thermally recoverable fractions, in particular, shredding and drying, which enables calorific value to increase to 22.0 MJ/kg. Comparative evaluation of different waste flow properties has revealed that the choice of MSW energy recovery scenario should be based on MSW composition and its thermal properties. In contrast to MSW incineration, the generation of high-calorie fuel from MSW requires significant waste treatment, reduces energy-recyclable waste mass and produces "undesirable" energy recycling flows which can be disposed of in landfill.

ACKNOWLEDGEMENT

The work was supported by Act 211 Government of the Russian Federation, contract № 02.A 03.21.0006.

REFERENCES

[1] Энергетическая утилизация твердых бытовых отходов. Энергетический бюллетень: 48. (2017). Аналитический центр при правительстве Российской Федерации. Online. http://ac.gov.ru/files/publication/a/13175.pdf. Accessed on: 8 Apr 2018. (In Russian)

[2] «Энергия из отходов». Online. https://w2e.ru/news/belyy-dym-iz-truby-chto-vmeste-s-parom-vybrosit-v-vozdukh-msz-v-osinovo. Accessed on: 4 Apr 2018. (In Russian)

[3] Velis, C. et al., Solid recovered fuel: Materials flow analysis and fuel property development during the mechanical processing of biodried waste. *Environmental Science & Technology*, **47**(6), pp. 2957–2965, 2013. DOI: 10.1021/es3021815.

[4] Rada, E.C. & Ragazzi, M., Selective collection as a pretreatment for indirect solid recovered fuel generation. *Waste Management*, **34**(2), pp. 291–297, 2014. DOI: 10.1016/j.wasman.2013.11.013.

[5] Zaccariello, L., Cremiato, R. & Mastellone, M.L., Evaluation of municipal solid waste management performance by material flow analysis: Theoretical approach and case study. *Waste Management & Research*, **33**(10), pp 871–885, 2015. DOI: 10.1177/0734242x15595284.

[6] Nasrullah, M., Vainikka, P., Hannula, J., Hurme, M. & Karki, J., Mass, energy and material balances of SRF production process. Part 3: Solid recovered fuel produced from municipal solid waste. *Waste Management & Research*, **33**(2), pp. 146–156, 2015. DOI: 10.1177/0734242x14563375.

[7] Nasrullah, M., Vainikka, P., Hannula, J., Hurme, M. & Oinas, P. Elemental balance of SRF production process: solid recovered fuel produced from municipal solid waste.

Waste Management & Research, **34**(1) pp. 38–46, 2016. DOI: 10.1177/0734242x15615697.

[8] Zhao, L. et al., Characterization of Singapore RDF resources and analysis of their heating value. *Sustainable Environment Research*, **26**(1), pp. 51–54, 2016. DOI: 10.1016/j.serj.2015.09.003.

[9] Zhou, H., Meng, A., Long, Y., Li Q. & Zhang, Y., Classification and comparison of municipal solid waste based on thermochemical characteristics. *Journal of the Air & Waste Management Association*, **64**(5), pp. 597–616, 2014. DOI: 10.1080/10962247.2013.873094.

[10] Rotter, V.S., Kost, Th., Winkler, J. & Bilitewski, B., Material flow analysis of RDF-production processes. *Waste Management*, **24**(10), pp. 1005–1021, 2004. DOI: 10.1016/j.wasman.2004.07.015.

[11] Ranieri, E., Ionescu, G., Fedele, A., Palmieri, E., Ranieri, A.C. & Campanaro, V., Sampling, characterisation and processing of solid recovered fuel production from municipal solid waste: An Italian plant case study. *Waste Management & Research*, **35**(8), pp. 890–898, 2017. DOI: 10.1177/0734242x17716276.

[12] Bushhinin, V.V., Lomtev, A.Yu., Budko, A.G. & Pakhtinov, V.M., Альтернативное топливо из твердых бытовых отходов. *Твердые бытовые отходы*, **4**, pp. 38–41, 2015 (in Russian).

[13] ИТС 9–2015, Информационно-технический справочник по наилучшим доступным технологиям. Обезвреживание отходов термическим способом (сжигание отходов). Бюро НДТ, 2015. (In Russian)

[14] Vaysberg, L.A., Mikhailova, N.V. & Gerasimov, A.M., Топливо для цементной промышленности из твердых коммунальных отходов. *Химия твердого топлива*, **1**, pp. 62–70, 2017 (in Russian).

[15] Bazyleva, Ya.V., Перспективы производства вторичного топлива из отходов. *Твердые бытовые отходы*, **11**, pp. 24–27, 2015 (in Russian).

[16] Krivoshein, V.G., Оценка энергетического потенциала ТБО на примере г. Перми. *Экология и промышленность России*, **1**, pp. 45–47, 2009 (in Russian).

[17] Ilyinikh, G.V., Slyusar, N.N. & Korotaev. VN., Морфологический состав отходов: основные тенденции изменения. *Твердые бытовые отходы*, **8**, pp. 38–41, 2011 (in Russian).

[18] Polygalov, S.V., Ilinykh, G.V. & Korotaev, V.N., Зависимость теплотехнических свойств твердых коммунальных отходов от компонентного и фракционного состава. *Теоретическая и прикладная экология*, **2**, pp. 70–78, 2018 (in Russian).

[19] Tugov, A.N., Исследование процессов и технологий энергетической утилизации бытовых отходов для разработки отечественной ТЭС на ТБО. Автореф. дис. …док. техн. наук, Москва: НИУ МЭИ, 2012 (in Russian).

[20] Brunner, P.H. & Rechberger, H., *Practical Handbook of Material flow Analysis*, CRC Press LLC, 2004.

[21] Lederer, J. et al., *Description Of The Environmental Pressures For Each Relevant Combination Of Material And Recycling/Treatment Technology*, Vienna University of Technology—Institute for Water Quality and Waste Management IWA: Austria, p. 176, 2009.

[22] Walter, RN., *Combustion and Incineration Processes*, Marcel Dekker, Inc., 2002.

[23] Komilis, D., Kissas, K. & Symeonidis, A., Effect of organic matter and moisture on the calorific value of solid wastes: An update of the Tanner diagram. *Waste Management*, **34**(2), pp. 249–255, 2014. DOI: 10.1016/j.wasman.2013.09.023.

[24] Lederer, J., Trinkel, V. & Fellner, J., Wide-scale utilization of MSWI fly ashes in cement production and its impact on average heavy metal contents in cements: The case of Austria. *Waste Management*, **60**, pp. 247–258, 2017. DOI: 10.1016/j.wasman.2016.10.022.
[25] Belousov, V.N., Smirnova, O.S. & Smorodin, S.N., Топливо и теория горения. СПб: ВШТЭ СПбГУПТД, 2016 (in Russian).

SPECIAL WASTE VALORIZATION AND RENEWABLE ENERGY GENERATION UNDER A CIRCULAR ECONOMY: WHICH PRIORITIES?

ELENA CRISTINA RADA
Department of Theoretical and Applied Sciences, University of Insubria, Italy

ABSTRACT

The European Union regulation for emissions from waste incinerators had a significant evolution in the last 20 years. That allowed reaching a clear improvement in the environmental performances of the plants, synchronized with specific requests of compliance of the more and more stringent limits set at regulatory level. Today the main questions for the sector in Europe seem mostly related to the role of waste to energy plants in a scenario of circular economy and to the role of those plants in the generation of energy only in part renewable. It seems that the topic of the local environmental and health impact has become out to date in European Union. In spite of that, a big question remains on the suitability of the present regulation for the future of the sector, including co-incineration by the use of Secondary Recovered Fuel from waste as substitute of coal in cement works. The present article analyzes an aspect that should be a priority higher than the concept of circular economy and energy generation. Indeed, a question remains on the criticalities that could be present in the European Union for under-estimation of the role of heavy metals, in spite of the evolution of the regulation. If we look at the experience of the University of Trento (where the Author worked in various research projects since 2003) the last decade has been characterized by recurrent warnings from that University about the excessive simplification of the regulation in case of heavy metals. As an example, in case of waste incineration, a set of heavy metals are managed by summing their concentrations to be compared as sum to the regulation limit: no valorization of the specific knowledge on toxicity is made. In order to point out the consequences of this simplification, a few case-studies demonstrating a potential criticality on the role of Cr^{VI} are discussed. A control methodology integrating the present one is discussed in this article, to propose a solution suitable to avoid under-estimations of human exposure to heavy metals from incineration and co-incineration. This approach is particularly important when special waste is combusted as the composition of the input can present heavy metals concentrations higher than municipal solid waste.

Keywords: Cr^{VI}, emissions, heavy metals, incineration, waste.

1 INTRODUCTION

The European Union (EU) regulation for emissions into the atmosphere from waste incinerators had a significant evolution in the last two decades. That allowed reaching a demonstrated improvement in the environmental performances of the plants, synchronized with specific progressive requests of compliance of limits set at regulatory level.

A significant case concerns PCDD/Fs emissions from the incinerator sector. The introduction of more stringent limits in the late 1990s allowed a decrease of three orders of magnitude of the emissions approved in Italy thanks to the introduction of the limit of 0.1 ng_{TEQ}/Nm^3 at the stack. The emission inventories confirmed a similar trend in EU. The same results (three orders of magnitude lower) did not characterize the sector of heavy metal emissions. The potential consequences are discussed in this article.

Nowadays, at local level, a few authorizations introduce the concept of guaranteed limits, more stringent than the authorized ones. In some cases, the difference between the regulation limits and the guaranteed ones is relevant. A case where this aspect can be seen extensively is the one of the incinerators of Bozen, Italy [1]. Here, differences are shown in Table 1.

WIT Transactions on Ecology and the Environment, Vol 222, © 2019 WIT Press
www.witpress.com, ISSN 1743-3541 (on-line)
doi:10.2495/EQ180141

Table 1: Comparison between regulation limits and guaranteed limits of pollutants emitted from the Bozen plant, receiving municipal and special waste [2].

		Regulation		Guaranteed		Reduction
Pollutants	Units	Aver.	Period	Aver.	Period	
Dust	mg/Nm^3	10	24 h	1.5	24 h	85%
HCl	mg/Nm^3	10	24 h	2	24 h	80%
HF	mg/Nm^3	1	24 h	0.25	24 h	75%
SO_2	mg/Nm^3	50	24 h	10	24 h	80%
NO_x	mg/Nm^3	200	24 h	40	24 h	80%
NH_3	mg/Nm^3	30	24 h	10	24 h	67%
Cd+Tl	mg/Nm^3	0.05	8 h	0.025	8 h	50%
Hg	mg/Nm^3	0.05	8 h	0.025	8 h	50%
Heavy metals	mg/Nm^3	0.5	1 h	0.25	1 h	50%
PCDD/F	ng_{TEQ}/Nm^3	0.1	8 h	0.025	8 h	75%

Possibly, these guaranteed limits are the most stringent in EU for the sector of incineration. What must be pointed out is that Table 1 is that the case of heavy metals is the less involved in the emission reduction targets. Some important consequences of that are discussed below in this article.

Today the main questions for the sector seem mostly related to the role of waste to energy plants in a scenario of circular economy and to the role of those plants in the generation of energy only in part renewable [3]–[5]. Indeed, key documents of waste to energy and circular economy [6] do not refer specifically to the issue of local impact on environment and health. More in details, in 2015, the European Commission proposed specific recommendations on the waste legislations in force, linked with Circular Economy (CE) [7]. The aim was "to stimulate Europe's transition towards a circular economy which will boost global competitiveness, foster sustainable economic growth and generate new jobs" [7]. The EU action program proposed measures involving the whole cycle: "from production and consumption to waste management and the market for secondary raw materials" [7]. The kind of specific key targets for all the member states were the following [7]:

- To get a recycling common target by 2030 for Municipal Solid Waste (MSW) and packaging waste;
- To reduce landfilling of MSW by 2030;
- To prohibit landfilling of waste selectively collected;
- To rationalize definitions, methods and standards for recycling rates all over EU;
- To have concrete forms to support the close-loop of the material re-use option;
- To stimulate EU towards a sustainable market place by specific actions.

Small modifications were introduced afterwards, without modifying the overall vision given by the previous document.

The above objectives relate to Municipal Solid Waste because European Union has considered that Circular Economy introduction in that field may also be driving for Special Waste (SW).

If we consider CE a priority in the EU waste management, the mentioned package of documents underlines the importance of reuse and recycling before energy recovery: the target is to close the loop of product lifecycles by an increase and an optimization of and re-use. This target is seen also in terms of favorable global balances for both the environment and the economy.

In the field of emissions from waste to energy plants, the discussion on BREFs (BAT Reference Report) is still active [8]. However, the approach is seen in terms of "conventional" emission improvements and not in terms of emission reduction targets. Indeed the control strategy at the stack does not change, in spite of the potential criticalities underlined below in this article.

The concept of partial renewable source for waste is intensively discussed in Italy every time a waste to energy plant is proposed. Recent criteria have been introduced in the sector allowing a choice between a detailed analysis of the biomass-like content of the waste and a flat approach: when the percentage of biodegradable matter in the waste fed to a plant is not known, that fraction can be assumed equal to 51%. That can be done under specific conditions described in the paragraph 6 of Annex 2 of Ministerial Decree of July 6th, 2012, modified by Law no. 221/2015 art. 24 [9].

In the same period, in Italy, no new decision concerned a different approach related to incineration emission management for new emission targets (thus for new health risk targets).

The health impact from waste incinerators seems a topic more interesting in the Chinese context [10]. If we look at the recent production of articles in the Scopus data base, most of the recent works on this topic are authored in China. It seems, in EU, that the reached level of knowledge on heath and incineration and the methodology of control are considered adequate in the scientific domains.

In spite of the reduced interest on the local impact on health in EU, a big question remains on the suitability of the regulation in force for the future of the sector, including co-incineration by the use of Secondary Recovered Fuel from waste as substitute of coal in cement works. Indeed an issue remains on the criticalities that could be present in the EU for under-estimation of the role of heavy metals, in spite of the past evolution of the regulation on emissions into the atmosphere. The present article focuses on it.

2 CRITICALITIES OF HEAVY METALS EMISSION MANAGEMENT

If we look at the experience of the University of Trento, where the Author worked in various research projects since 2003, the last decade has been characterized by recurrent reporting to local Authorities (e.g. Environmental Protection Agencies) about the excessive simplification of the regulation in case of heavy metals. As an example, in case of waste incineration, heavy metals are generally managed by summing their concentrations to be compared as sum to the regulation limit: no valorization of the specific knowledge on toxicity is made, apart from Hg, considered alone. This concept will be analyzed in details in the following case studies. The direct consequence of this simplification is the lack of valorization of the detailed knowledge on the health effects of heavy metals, thanks to the progress in toxicology resumed below.

Differently from conventional macro-contaminants (as NO_2, SO_2, CO and some VOCs), for other pollutants the pathways of indirect exposure cause not negligible impacts (even dominants for a few compounds). This is the case of the toxic and persistent atmospheric pollutants, which include heavy metals, PCDD/Fs, PCBs and PAHs. Their characteristics of persistence and cumulability in the environment require a more articulated approach to calculate the risk for health associated to the human exposure. From the methodologic point of view, indeed, the study of their impact on health must be based on the concept of multiple pathways responsible of effects on the individuals exposed.

Pollutants, after emitted into the atmosphere, can penetrate into the human body directly (through inhalation) or can reach the soil by deposition. Here three pathways can give indirect exposure:

- Accidental ingestion of soil contaminated by the atmospheric depositions of pollutants;
- Dermal contact with contaminated soil;
- Ingestion of food contaminated by the pollutant's depositions to soil, leaves, fruits.

For many toxic and persistent atmospheric pollutants, e.g. PCDD/F, PCB and heavy metals, food chain is for humans an important way of risk, when not the main one. As consequence, among the pollutants of concern for the simulations of dispersion into the atmosphere, thus interesting for the health risk assessment, there are many carcinogenic pollutants. The International Agency for Research on Cancer, IARC, classified a number of compounds on the base of the outcomes of epidemiologic studies and experiments on animals, on the reliability and replicability of results [11]. In particular, IARC subdivides the carcinogenic compounds in four groups:

- Group 1: Carcinogenic to humans
- Group 2A: Probably carcinogenic to humans
- Group 2B: Possibly carcinogenic to humans
- Group 3: Unclassifiable as to carcinogenicity in humans
- Group 4: Probably not carcinogenic to humans

Table 2 reports the compounds typically used for the assessment of the health risk in case of incinerators and their IARC group. In order to simplify, in the present article only the carcinogenic risk is considered as it is sufficient to demonstrate the potential underestimation of the effects of heavy metals with the regulation in force.

Data on toxicity of each compound are available in form of Cancer Slope Factor (SF) for the carcinogenic substances. SF, expressed in (mg kg_{bw}^{-1} $giorno^{-1})^{-1}$ where "bw" means body

Table 2: Classification from IARC for the carcinogenic compounds subject of the health risk assessment [11].

Compound	Group IARC
Arsenic	1
Cadmium	1
Chromium VI	1
Mercury	3 and 2B[a]
Nickel	1
PAHs	from 1 to 3
PCDD/Fs	1
PCB dioxin-like	1

[a] For elemental Mercury and Methilmercury respectively.

Table 3: Toxicologic characteristics of the considered pollutants.

	SF_{inhal} $[(mg\ kg_{bw}^{-1}\ d^{-1})^{-1}]$	SF_{oral} $[(mg\ kg_{bw}^{-1}\ d^{-1})^{-1}]$
Cadmium	6.30E+00	-
Tallium	-	-
Arsenic	1.50E+00	1.50E+00
Chromium III	-	-
Chromium VI	2.94E+02	-
Nickel	9.10E-01	9.10E-01
PCDD/F	1.33E+05	1.33E+05
PAH(*)	4.20E+00	7.30E+00
PCB dl	1.33E+04	1.33E+04

(*) SFs of the most toxic PAHs are reported.

weight is defined as the incremental cancer risk associated with a unit dose of a carcinogen. It is the slope of the curve representing the relationship between dose and cancer risk. Depending on the exposure pathway, inhalation or ingestion, we distinguish between SF_{inhal} and SF_{oral}. Table 3 reports the toxicologic parameters available for the considered pollutants.

Rows for Tallium and Chromium III are reported even if they are blank, in order to take into account what discussed below.

From Table 3 we can remark that:

- Heavy metals show SFs in a range of two orders of magnitude
- In spite of that, the regulation in force sums heavy metal concentrations with no regard to their specific toxicity; as examples:
    - In the regulation for incineration emissions at the stack, Cd+Tl are summed even if only one has a SF
    - Arsenic, Chromium and Nickel are summed with other six heavy metals even if only three have a SF
- The approach adopted for PCDD/Fs and PCBs, based on an equivalent toxicity, is not used in the sector of heavy metals

Possibly one reason of this simplified approach comes from the fact that in the past the organic micropollutants like PCDD/Fs drove the value of the total risk from an incinerator. As demonstrated in the following case studies related to the Cr^{VI} issue, this cannot be longer accepted without an update of the criteria control as:

- The significant reduction of operating values in the emissions of PCDD/Fs changed the role of dominant pollutant; indeed in some cases heavy metals emerge driving the total risk value
- The approach in force in terms of limits for heavy metals based on a raw sum of concentration values of nine of them, allows critical cases when, for instance, Cr^{VI} is a characteristic and not negligible pollutant in the emissions.

3 THE CHROMIUM ISSUE

In order to point out the consequences of the simplification in the management of the issue of heavy metals emission, a few case studies demonstrating a potential criticality on the role of Cr^{VI} are discussed. A control methodology integrating the present one is discussed too, to set a solution suitable to avoid under-estimations of human exposure. This approach is particularly important when special waste is combusted as the composition of the input can present heavy metals concentrations higher than RMSW (residual MSW, i.e. the one remaining after selective collection).

3.1 RMSW and special waste incinerator's case study

Supposing to consider a limit value of Cd at the stack as 50 μg/Nm^3, it is clear that also for Cr^{VI}, carcinogenic, the emission at the stack must be coherent with that request. Taking into account the different SF of Cr^{VI}, along the pathway adopted to authorize the emissions of an Italian plant, it was proposed to introduce a maximum value of Cr^{VI} to be guaranteed at the stack as 1 μg/Nm^3 calculated on yearly basis (being a carcinogenic compound, its relevance is on a long period of exposure) [12].

Considering the novelty of the approach, a proposal of criteria for measuring it on yearly basis was left to the plant proposer.

As a consequence, the list of guaranteed values, integrating or partially substituting the regulation in force in that period, resulted as Table 4. This Table is valid for solutions of

Table 4: Concentration values to guarantee at the stack as integration or partial substitution of the regulation in force (values at 11% O_2, dry off-gas). The parameters that do not change from the regulation are not reported.

Parameter	Values to be guaranteed
Dust (daily average)	2 mg/Nm^3
TOC (daily average)	5 mg/Nm^3
HCl (daily average)	2 mg/Nm^3
HF (daily average)	0.5 mg/Nm^3
NO_x (daily average)	40 mg/Nm^3
NO_x (hourly average)	200 mg/Nm^3
SO_2 (daily average)	25 mg/Nm^3
NH_3 (daily average)	5 mg/Nm^3
Cd+Tl (hourly average)	0.01 mg/Nm^3
Hg (hourly average)	0.02 mg/Nm^3
Sb+As+Pb+Cr+Co+Cu+Mn+Ni+V (hourly average)	0.2 mg/Nm^3
PCDD/F (yearly average)	0.03 ng_{TEQ}/Nm^3
BaP (yearly average)	0.1 μg/Nm^3
Cr^{VI} (yearly average)	1 μg/Nm^3
Other parameters	The eight hour sampling for PCDD/F control and its limit of 0.1 ng_{TEQ}/Nm^3 was confirmed (it is not reported in this Table); the one proposed on yearly basis is an integration of it; the same integration can be seen with BaP, whilst other parameters did not change (e.g. CO)

conventional combustion and must be completed with other parameters set by the regulation (e.g.: CO).

It is an approach that even 10 years ago recognized the limits of the regulation in force but did not focused in detail on the matter of the equivalent impact given by different carcinogenic compounds.

3.2 Tannery sludge thermochemical treatment case study

Two big leather-tanning industrial areas are present in Italy. Since long time, the attention was put on the role of Chromium that is used as Cr^{III} in the tanning activities, because:

- It can be found in trace in the leather product, causing allergic reactions; that limits also the possibility of the reuse of the material in the frame of a circular economy;
- It is transformed into Cr^{VI} if the final treatment of the generated sludge (from wastewater treatment plants) is based on oxidative conditions; that limits the energy recovery from this sludge in spite of the fact that is assimilated to a discarded biomass, thus a renewable source of energy.

In order to find a solution, local and national Authorities promoted a lot of research on the topic in Italy. What emerges to whomever analyzes the problem is an inadequate regulation on emissions to air, as the high content of Cr^{III} in sludge could cause high concentrations of Cr^{VI} at the stack in case of combustion, with no problems of complying the national regulation on heavy metals emissions. Just to have an example, a concentration of Cr^{VI} equal to 100 $\mu g/Nm^3$ and the rest of the heavy metals equal to 400 $\mu g/Nm^3$ results acceptable from the regulation even if the health risk only from Cr^{VI} results significantly higher than the risk from PCDD/F assumed equal to the limit (0.1 ng_{TEQ}/Nm^3).

The problem can partially emerge also in the case of gasification, as a local formation of Cr^{VI} can be found for local oxidizing conditions even if the global conditions are sub-stoichiometric.

Pyrolysis is worst: the combustion of the produced char offers to Cr^{III} there contained the possibility to be oxidized to Cr^{VI}.

At the end, an important solution is prevention: tanning without Cr (wet white leather) is more and more introduced in the sector; however, a long time will need to get a full conversion of the approach because of a higher cost compared to the use of Cr^{III}. Thus, an integration of the regulatory approach is compulsory.

3.3 Woody waste combustion case study

In the latest years, EU has started a review of the regulations about the combustion of woody biomass. In Italy virgin wood falls in the category of fuels [13]. Treated wood deriving from wood-furniture industry rejects is considered as waste with consequences on the option of combustion: its combustion is subjected to more stringent restrictions in term of process monitoring and emission limit. A question raised recently in Italy about the possibility to classify slightly treated wood as fuel. A research was developed and analyses before (input) and after pilot combustions (outputs) were made, dealing with organic micropollutants [14] and heavy metals (but not detailing on Cr^{VI}). What is interesting for the present article is the analysis of the heavy metals content in virgin wood. In practice, if wood comes from polluted ground, an anomalous heavy metals amount (Cr included) can enter a burner without any control of its emissions. This problem, not yet faced with in Italy, is clear in Germany where some limits are set. In this case, the approach of Germany should become a reference for other countries where no control on wood composition is requested. However, the problem remains in case treated wood is burnt as its Cr content usually is not characterized.

3.4 Special waste gasification case study

A recent proposal for special waste gasification (followed by immediate combustion of the syngas) has been put forward in the North of Italy [15]. The plant could treat 95,000 t/y of waste with a LHV as 16 MJ/kg. Higher values of LHV are acceptable reducing the feed.

The technology allows material recovery thanks to the integrated vetrification of bottom ash.

Also in this case, the authorization pathway suffers from the aggregation of heavy metals in the limit values to be complied with:

- The sum of Cd and Tl has a limit
- The sum of Antimonium, Arsenic, Lead, total Chromium, Cobalt, Manganese, Nickel e Vanadium (Sb + As + Cr + Pb + Co + Cu + Mn + Ni + V) has a limit

Being a proposal of plant, it was not possible to know the speciation of metals inside these two categories. In this case study, the speciation was taken from a report of an Italian Region on the monitoring of emissions from modern incinerators [16]. From the average values of the sampling of each metal the fractionation of heavy metals was obtained for the two categories (Table 5). By this way, when facing with dispersion modeling, for each cell of the domain of calculation it was possible to assess the concentrations of specific heavy metals starting from the total concentration of each category.

As the regulation does not set a specific emission limit for Cr^{VI}, in order to develop the health risk analysis, an assumption of a ratio between Cr^{VI} and total Cr was taken.

As precautionary assumption, data from the Environmental Agency of United Kingdom were taken, based on analyses of flyash separated by the off-gas treatment lines of ten incinerators in that country among 2007 and 2009 [17]. The highest ratio resulted 2.1%, ratio between Cr^{VI} and total Cr. By this way the health risk (cancer risk) can be developed based on the role of Cr^{VI}. Results demonstrated its comparative relevance against PCDD/Fs [15]. Indeed, low values of dioxin were guaranteed in that proposal thanks to the ability of gasification in inhibiting their formation. Thus, the prevailing carcinogenic driver resulted in

Table 5: Contributions related to each heavy metal on the total of the two categories, whose emissions are regulated.

Heavy metals	Mass fraction on the total [-]
Cd	0.907
Tl	0.093
TOTAL	1.000
Ni	0.034
Mn	0.058
Cr total	0.684
Pb	0.075
Sb	0.109
V	0.034
As	0.003
Co	0.003
TOTAL	1.000

www.witpress.com, ISSN 1743-3541 (on-line)

that case Cr^{VI} in a context where the national regulation does not take into account this possibility.

In this case study, the renewability of special waste was analyzed in detail showing that the more virtuous case should be the analytic assessment of the biomass-like characteristics of the waste. Flat assumptions on the biogenic characteristics of special waste can be far from the real context, considering that the feed can vary from cellulosic based matter to plastic based matter.

Another interesting aspect of this case study concerns mass valorization of part of the feed: the integration of vitrification is coupled with a partial adjustable recycling of flyash in the vitrification zone, in order to favor the trapping of heavy metals and other inert parts. By this way, flyash will contribute to the generation of a product (the vitrified output). This pathway can be considered a CE oriented downcycling.

Moreover, the off-gas treatment line will adopt a double step of filtration based on the concept of a preliminary filtration subsidiary to a second step where Sodium Bicarbonate is added and flyash is sent to a centralized recycling plant for the regeneration of the additive that can be used back in the plant. This is an option coherent with the principles of CE.

3.5 SRF and sewage sludge combustion case study

The sector of SRF (Secondary Refused Fuel) is not exempt from the potential problems from Cr^{VI}. That is demonstrated by some cases in the sector of cement works authorized for co-combustion of special waste. An interesting example is the one of an Italian cement factory [18] where SRF and thermally dried sewage sludge are authorized for co-combustion. An attention to the problem is demonstrated by the fact that those auxiliary fuels can be accepted at the gate of the plant if they comply with limits of some pollutants including total Chromium and Cr^{VI} (Tables 6 and 7).

The use of SRF and sludge is respectively a partial and total substitution of fossil fuel (coal) thus it is an activity of renewable energy exploitation. Moreover, it is related also to circular economy, in the form of downcycling (the valorized mass is included as ash in the final product, the cement). That the whole approach seems to be enhanced. In reality, what lacks in the approach is a control of the situation at the stack, because in that case the control criteria are based only on the verification of a limit as sum of some heavy metals, as usual. Table 8 shows the limits to comply with.

Table 6: Characteristics required for SRF admitted to the cement factory.

Parameters	Limits	Value
LHV	Minimum, as is	15 MJ/kg
Moisture	Maximum	25%
Cl	Referred to dry matter (DM)	≤0.9%
S	Referred to dry matter	≤0.6%
Ash	Referred to dry matter	≤20%
Pb	Volatile, mg/kg referred to DM	≤200
Cu	Soluble compounds, mg/kg referred to DM	≤300
Mn	mg/kg referred to DM	≤400
Ni	mg/kg referred to DM	≤40
As	mg/kg referred to DM	≤9
Cd+Hg	mg/kg referred to DM	≤7
Cr	mg/kg referred to DM	≤100

Table 7: Characteristics of thermally dried sewage sludge admitted to the cement factory.

Parameters	Limits	Value
LHV	Minimum, as is	7 MJ/kg
Moisture	Maximum	15%
P	Referred to dry matter (DM)	≤3%
Ash	Referred to DM	≤55%
Cr^{VI}	mg/kg, referred to DM	≤10
Cd+Tl+Hg	mg/kg, referred to DM	≤10

Table 8: Limits for heavy metals at the stack (average daily values).

Parameters	Units	Value (limit)
Hg	mg/Nm^3	0.05
Σ Cd+Tl	mg/Nm^3	0.05
Σ Sb + As + Cr + Pb + Co + Cu + Mn + Ni + V	mg/Nm^3	0.5

3.6 A proposal for enhanced Cr^{VI} control

What pointed out for Cr^{VI} is not limited only to the sector of special waste. Painting activities at industrial level, printing processes and steel making plants can have the same anomalies in the authorization pathway (in Italy and possibly abroad).

For example, painting is assumed often to be a small activity thus a simplification of the authorization is typical of the approach. No Environmental Impact Assessment is necessary in Italy under a threshold of production and a simplified control of the emitted streams is the only request. In the case of [15] it was demonstrated that closing a painting plant to give room to a modern gasifier would allow a reduction of the local impact of Cr^{VI} (referring to the authorized values). That because small activities are authorized to emit at low height with a consequent (often not estimated) reduction of the dilution capability of the pollutants released. Thus, the proposal explained below must be considered also in contexts out of the waste sector.

Taking into account what found in the discussed case studies, an enhanced approach to take Cr^{VI} impact under control could be the following:

- Ask for total Cr values at the stack from the laboratories involved in emission characterization (avoiding receiving only sum of multiple heavy metals)
- Set a preliminary threshold for total Cr (taking into account comparatively the toxicity of the other pollutants emitted and the typical ratio Cr^{VI}/total Cr in the sector) so that, when total Cr analyses result beyond it:
    - specific analyses of Cr^{VI} will be performed (case specific) in addition to total Cr analyses
    - a risk analysis will be developed in order verify if a different threshold for total Cr emissions must be introduced

It is clear that in case the plant is already authorized, the collaboration of the company managing it is compulsory. Moreover, the focus on Cr^{VI} shown in this article should be extended to other heavy metals.

Table 9: CE actions, renewable energy generation aspects and enhanced visions of health impact minimization in the sector of waste to energy (from the case studies).

Type	*Description*	*Priority*
Circular economy	Additive regeneration from flyash	It depends on new plants construction
Circular economy	Flyash integrated vitrification	It depends on new plants construction
Circular economy	Bottom ash vitrification	It depends on new plants construction
Renewable energy	Avoid flat approaches and promote detailed analyses of biogenic content in the waste	It depends on new plants construction, as the approach can be changed only for new authorizations
Health risk	Enhanced Cr^{VI} control	Immediate
Health risk	Enhanced approach to set heavy metals emission limits coherent with different levels of toxicity	High priority to develop a method based on the concept of equivalence toxicity (similarly to the one for PCDD/F)

4 PRIORITIES

Table 9 resumes the aspects of interest found in the presented case studies concerning CE actions, renewable energy generation and enhanced visions of health impact minimization in the frame of waste to energy.

The case of bottom ash recycling in centralized plants must be cited to complete the vision (this option can be adopted also for existing plants).

As shown in Table 9, a part of the interventions depends on new plants construction. The documents issued by EU on CE seems to expect a strong limitation of new plants. In reality, the present availability of plants for special waste is not homogeneous in Europe and/or is not sufficient. The case of Italy is clear. We see presently an important export of special waste towards other European countries against the principle of proximity in treating waste. In parallel, the growing difficulty for waste export, specifically source separated plastic of low quality, is making critical this situation. The demonstration of this crisis is the anomalous series of recent fire events in Italian companies storing plastics from selective collection.

5 CONCLUSIONS

This paper pointed out some criticalities that remain in the sector of emission control and regulation in EU concerning heavy metals. After focusing on the case of Cr^{VI}, it is clear that the approach, presently adopted according to the regulation in force, does not valorize the scientific knowledge available on the toxicity of heavy metals, exposing the population to potential risks that are not assessed adequately in details.

A proposal to manage the case of Cr^{VI} has been put forward in order to integrate the present regulation and avoid not optimized extra-costs.

The criticalities pointed out in this article may be more relevant when the sector of special waste is taken into account: the heavy metals content in the input of a thermo-chemical plant can be more variable than the case of MSW and potentially riskier.

Considering the relevance of the criticality related to heavy metals emission authorization, the Author is working on a novel approach that should take into account the concept of equivalent toxicity of the different heavy metals.

The modification of the approach in managing heavy metals control is a priority compared to the options analyzed that could be adopted in the waste to energy sector to make it more coherent with CE and sustainability principles.

REFERENCES

[1] Ragazzi, M., Tirler, W., Angelucci, G., Zardi, D. & Rada, E.C., Management of atmospheric pollutants from waste incineration processes: The case of Bozen. *Waste Management & Research*, **31**(3), pp. 235–240, 2013. DOI: 10.1177/0734242x12472707.

[2] www.provinz.bz.it/service/resdownload.aspx?source=VIA-UVP&ID=F01C7619D29C2838E040007F01003CC6. Accessed on: 24 May 2018.

[3] Rada, E.C., Cioca, L.I. & Ionescu, G., Energy recovery from municipal solid waste in EU: Proposals to assess the management performance under a circular economy perspective. *MATEC Web of Conferences,* **121**, p. 05006.

[4] Rada, E.C., Ragazzi, M., Torretta, V., Castagna, G., Adami, L. & Cioca, L.I., Circular economy and waste to energy. *AIP Conference Proceedings,* **1968**, p. 030050, 2018.

[5] Malinauskaite, J. et al., Municipal solid waste management and waste-to-energy in the context of a circular economy and energy recycling in Europe. *Energy,* **141**, pp. 2013–2044, 2017. DOI: 10.1016/j.energy.2017.11.128.

[6] EU, 2017. Communication from the Commission to the European Parliament, the Council, the European Economic and Social Committee and the Committee of the Regions—The role of waste-to-energy in the circular economy—Brussels, 26.1.2017 COM(2017) 34 final.

[7] http://ec.europa.eu/environment/circular-economy/. Accessed on: 24 May 2018.

[8] http://eur-lex.europa.eu/legal-content/IT/TXT/PDF/?uri=CONSIL:ST_7274_2017_INIT&qid=1491979061848&from=IT. Accessed on: 24 May 2018.

[9] http://eippcb.jrc.ec.europa.eu/reference/BREF/WI/WI_5_24-05-2017_web.pdf. Accessed on: 24 May 2018.

[10] www.gazzettaufficiale.it/eli/id/2016/1/18/16G00006/sg. Accessed on: 24 May 2018 (Law Dec. 28 2015, no.221, "Disposizioni in materia ambientale per promuovere misure di green economy e per il contenimento dell'uso eccessivo di risorse naturali").

[11] Lu, J.W., Zhang, S., Hai, J. & Lei, M., Status and perspectives of municipal solid waste incineration in China: A comparison with developed regions. *Waste Management*, **69**, pp. 170–186, 2017. DOI: 10.1016/j.wasman.2017.04.014.

[12] www.iarc.fr/. Accessed on: 24 May 2018.

[13] www.comune.trento.it/Aree-tematiche/Ambiente-e-territorio/Rifiuti-urbani. (Studio di Fattibilità - Concessione di lavori per la progettazione, realizzazione e gestione dell'impianto di combustione o altro trattamento termico con recupero energetico per rifiuti urbani e speciali assimilabili in località Ischia Podetti, nel Comune di Trento, Authonomous Province of Trento, June 2009).

[14] Girelli, E., Ragazzi, M., Malloci, E., Rada, E.C. & Paternoster, L., Agricultural biomass availability for energy conversion in Italy. *UPB Scientific Bulletin, Series C,* **74**(1), pp. 11–18, 2012.

[15] Passamani, G., Rada, E.C., Tirler, W., Tava, M., Torretta, V. & Ragazzi, M., PCDD/F emissions from virgin and treated wood combustion. *International Journal of Energy Production and Management,* **2**(1), pp. 17–27, 2017. DOI: 10.2495/eq-v2-n1-17-27.

[16] Rada, E.C., Castagna, G., Adami, L., Torretta, V. & Ragazzi, M., Compensation opportunities and waste-to-energy plants. *AIP Conference Proceedings*, **1968**, p. 030065, 2018.

[17] www.arpae.it/cms3/documenti/moniter/quaderni/03_emissioni.pdf. Accessed on: 24 May 2018.

[18] www.gov.uk/government/publications/waste-incinerators-guidance-on-impact-assessment-for-group-3-metals-stack. Accessed on: 24 May 2018.

[19] www.provincia.como.it/export/sites/provincia-como/temi/territorio/ambiente/autorizzazione-integrata-ambientale/autorizzazioni/holcim_3ries.pdf. Accessed on: 24 May 2018.

www.witpress.com, ISSN 1743-3541 (on-line)

Author index

www.ingramcontent.com/pod-product-compliance
Lightning Source LLC
LaVergne TN
LVHW062240200926
848484LV00006B/1578